Storey's Barn Guide **TO**
HORSE HANDLING
AND GROOMING

TEXT BY CHARNI LEWIS

Illustrations by Alison Schroeer

Storey Publishing

The mission of Storey Publishing is to serve our customers by publishing practical information that encourages personal independence in harmony with the environment.

Edited by Deborah Burns and Sarah Guare

Cover design by Vicky Vaughn Design

Text design and production by Jessica Armstrong

Cover photographs © Rachael Waller Photography (front); © Kevin Kennefick (back)

Interior photographs © Rachael Waller Photography: iv, 12, 28, 46, 70, 86, 104, 118, 120; © Phil Cardamone/iStockphoto.com: 34

Illustrations © Alison Schroeer

Printed in Hong Kong by Elegance

10 9 8 7 6 5 4 3 2 1

LIBRARY OF CONGRESS CATALOGING-IN-PUBLICATION DATA

Lewis, Charni.
    Storey's barn guide to horse handling and grooming / by Charni Lewis.
        p. cm.
    Includes index.
    ISBN 978-1-58017-657-6 (concealed wire-o : alk. paper)
    1. Horses—Handling. 2. Horses—Grooming.
    I. Title.
SF285.6.L49 2007
636.1'083—dc22

ACKNOWLEDGMENTS

I would like to thank:

Richard Catt, Matthew Lewis, and my family for their support and patience in completing this project.

Jan Rodriguez and Deb Burns for providing me with the opportunity to write this book.

Flintridge Riding Club, for providing such a great facility, and its members, for volunteering their time.

Phyllis Lambert, Wendy Averill, and Desdy Baggott for giving me such a good start in working with horses.

All of those horse people who have generously shared their time and knowledge with me.

CONTENTS

TIPS FOR APPROACHING

THE WAY YOU APPROACH YOUR HORSE WHEN YOU ENTER THE STALL OR CORRAL can set the tone for your whole ride. Enter with a relaxed but confident posture. This will show the horse that you are to be respected and will let him know that you are an ally, not a predator. Follow these tips when approaching.

1 DO NOT STARE DIRECTLY INTO YOUR HORSE'S EYE WHEN TRYING TO CATCH HIM. If you do, he may consider you a predator.

2 MAKE SURE THAT YOUR HALTER AND LEAD ROPE ARE ORGANIZED in such a way that they will not entangle you when you are trying to catch and halter your horse.

3 NEVER APPROACH A LOOSE HORSE FROM BEHIND. From a distance, ask him to move by making a "clucking" sound until he turns his head and shoulder toward you.

4 SUDDEN MOVEMENTS CAN SCARE YOUR HORSE, so move slowly and deliberately when approaching him.

5 ALWAYS BE WATCHFUL OF THE OTHER HORSES IN THE AREA. If you are trying to catch one horse that is loose with several other horses, make sure that you are not in the way of two horses having a disagreement.

6 IF THERE ARE OTHER HORSES NEARBY, RESIST THE URGE TO BRING A TREAT INTO THE CORRAL. The other horses may aggressively compete with each other to get the food and you could unintentionally land between two quarreling horses.

7 MAKE SURE TO CLOSE THE GATE BEHIND YOU OR STAY BETWEEN YOUR HORSE AND THE GATE so your horse does not escape before you catch him.

Appropriate Clothing

YOUR STYLE OF RIDING WILL INFLUENCE your choice of pants and boots. Tall English boots or half chaps with britches are comfortable for riding in an English saddle because they protect the rider's legs from getting pinched by the saddle's stirrup leathers. Western saddles have fenders to protect the rider's legs from stirrup straps so jeans and cowboy boots are comfortable for Western riding.

SAFETY TIPS

- Always wear a helmet when riding.

- Be careful when removing your jacket while in the saddle. Some horses will spook when you do this, and you may not be in a good position to grab the reins.

- Closed-toe shoes are required for safety.

- Hard-toed shoes such as boots are recommended when working around horses. They give protection in case a horse should step on your foot.

- A boot with a heel is required for riding to prevent the foot from slipping through the stirrup.

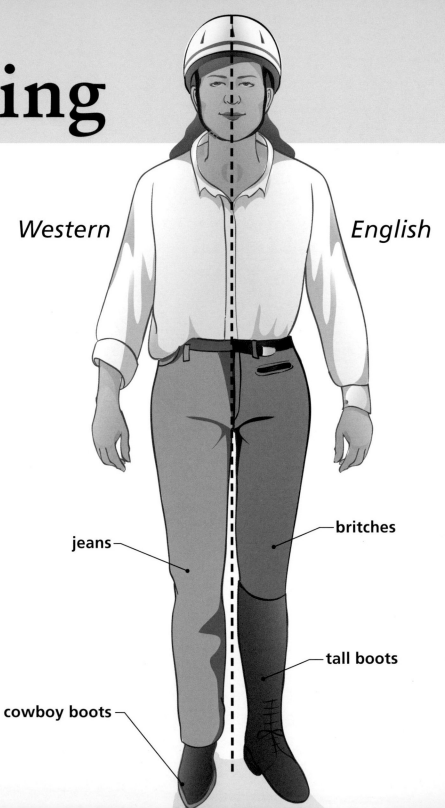

Western *English*

jeans

britches

tall boots

cowboy boots

FULL CHAPS protect legs from rubbing on the saddle and from external elements like trail brush.

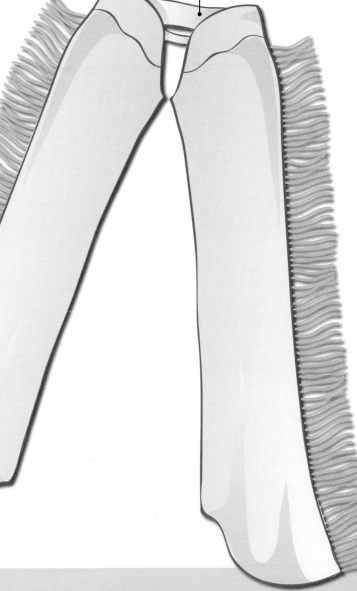

HALF CHAPS and **PADDOCK BOOTS** are an alternative to tall boots.

PROTECTION FROM THE ELEMENTS

- Thin or lined gloves, depending on the weather, are useful for handling reins or a lead rope.

- Use supportive undergarments like jockey shorts and sports bras for greater comfort while riding.

- Sunscreen is always a good idea when working outside.

- In cold weather wear layers while riding so that you can shed them as you warm up.

Moving Around Your Horse

BE CAUTIOUS WHEN WORKING BEHIND OR IN FRONT OF THE HORSE.
You should do all of your handling from the "safe zones" to the sides of his body. When moving around the horse, keep a relaxed body posture because horses interpret this as soothing and nonthreatening.

WALKING BEHIND AN UNTIED HORSE

If there is room or if the horse is not tied, walk in a relaxed, casual manner at least 8–10 feet away from his back to stay out of "kick range."

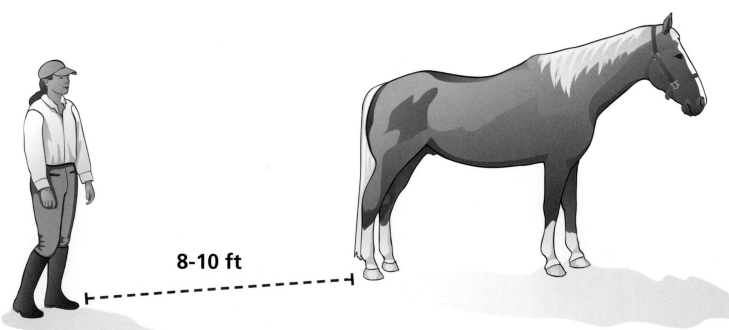

8-10 ft

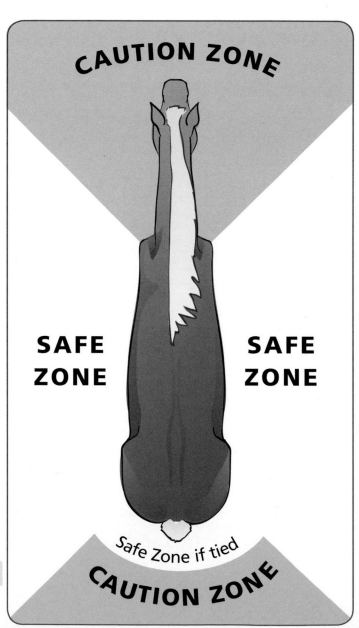

CAUTION ZONE

SAFE ZONE

SAFE ZONE

Safe Zone if tied

CAUTION ZONE

TIPS FOR WALKING NEAR TIED HORSES

- Keep a relaxed posture and gently outstretched hand when approaching and moving around the horse.

- Move slowly and methodically around your horse while speaking in a low to normal tone of voice. Horses are prey animals that startle easily.

- Be alert at all times. Any sudden movement from a person or an object such as a plastic bag will frighten a horse, causing him to react violently.

- Avoid placing yourself tightly between your horse and a solid object. If the horse is too close to a wall, use firm pressure with your hand on the horse's shoulder or hip and make a "cluck" sound to ask the horse to move over.

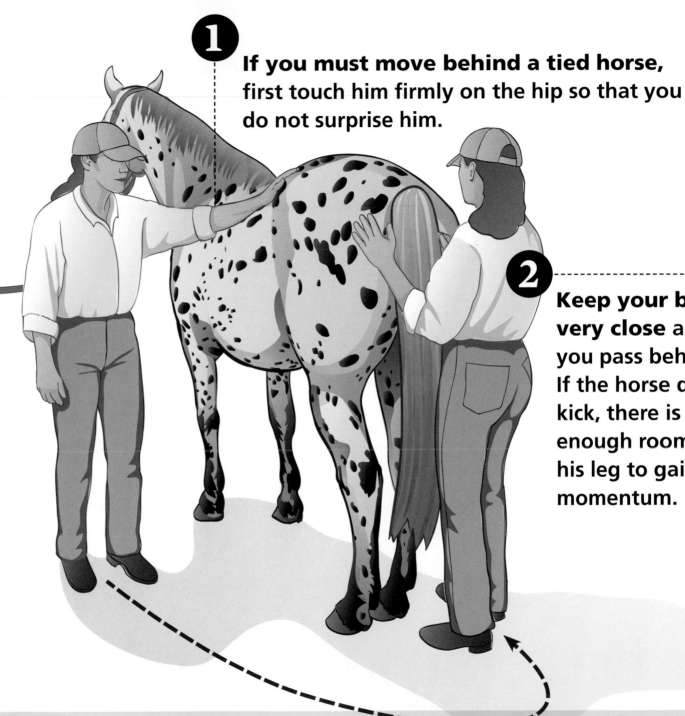

1 If you must move behind a tied horse, first touch him firmly on the hip so that you do not surprise him.

2 Keep your body very close as you pass behind. If the horse does kick, there is not enough room for his leg to gain momentum.

Catching a Horse in the Stall

Tools you will need: *halter, lead rope*

WHEN CATCHING A HORSE, relax your posture and don't make direct eye contact with him. Staring directly into a horse's eye is interpreted as an aggressive, predatory posture that will make him anxious and difficult to catch. Position the horse in the stall before you enter by "clucking" or speaking softly so that he is facing the front of the stall or has moved to allow you room to enter safely.

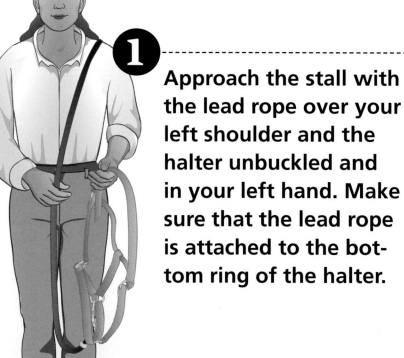

❶ Approach the stall with the lead rope over your left shoulder and the halter unbuckled and in your left hand. Make sure that the lead rope is attached to the bottom ring of the halter.

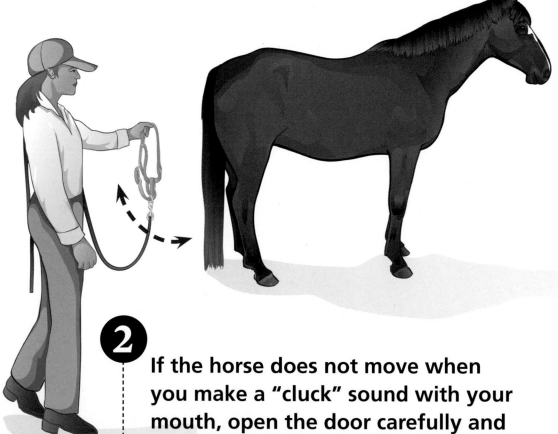

❷ If the horse does not move when you make a "cluck" sound with your mouth, open the door carefully and gently swing the halter in the direction of his haunches without touching him, so that he turns toward you.

3 Approach the horse's left side with a casually outstretched arm. If he will only present the right, approach the shoulder that is available. Touch his shoulder with a firm, reassuring hand.

4 Carefully put the loose end of the lead rope over the top of the horse's neck and slide the rope over until you can see it appear under the neck on the far side.

5 Grasp the end of the rope under the neck and pull it toward you so that you now have a loop of rope around the horse's neck.

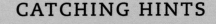

6 Move the loop toward the horse's head to control the horse until you get the halter on. If you are on his right side, hold the loop and carefully move around the front to the left side to halter the horse.

CATCHING HINTS

- Typically, it is best to handle a horse from the left side, because this is often how he is most used to being handled.

- Never wrap the lead rope around any part of your body.

- Always stay between the horse and the open stall door to prevent the horse from escaping and to prevent you from being trapped in the back of the stall.

Catching a Horse in a Corral

TO CATCH A HORSE SAFELY, walk at a normal pace, maintain a relaxed posture, and approach casually, without making eye contact. Don't forget to close the gate behind you.

HELPFUL TIPS

- If the horse runs away when you enter the corral, stand quietly in the middle of the corral and wait for him to stop.

- Walking in the direction of his head will cause the horse to slow down or turn. Walking in the direction of his haunches will make him move forward.

- It is best to approach the horse from the left side.

IF THE HORSE BEGINS TO MOVE, walk in a line that will intersect with his line of travel close to the edge of the corral.

IF THE HORSE'S HAUNCHES ARE FACING YOU, stand about 20 feet away and cluck to him while gently swinging the halter. This will prompt him to move his body parallel to the fence.

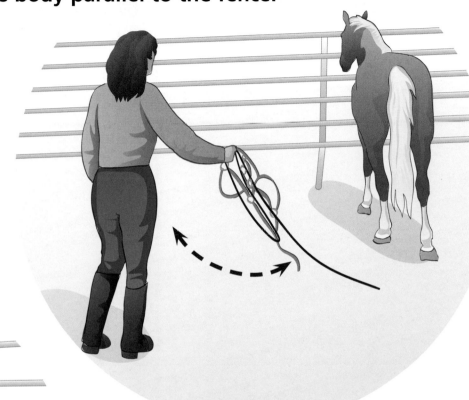

IF THE HORSE'S RIGHT SIDE IS FACING YOU, approach his right shoulder with a casually outstretched arm.

IF THE HORSE IS EXCEPTIONALLY HARD TO CATCH, look away from him as you approach and keep your shoulder turned away for the last 15 feet. Carefully place your hand on his shoulder.

Haltering a Horse

Tools you will need: *halter, lead rope*

USE A HALTER TO STEER AND STOP the horse when leading him. Make sure that the halter is the proper size and that the lead rope is attached to the bottom ring.

1 With a rope around the horse's neck, stand at his left side between his shoulder and head. Hold the halter buckle in your left hand and the halter strap in your right.

2 Reach under the horse's neck with the right hand, holding the halter strap in it, and pass the strap around his neck.

HALTERING RULES

- Keep your face well back from the horse's head while haltering so that if he flings his head it does not hit you.

- Never put your fingers inside of the rings of the halter or your hands between the horse and the halter.

- Never lead a horse with a halter alone; always use a lead rope.

- Never let the lead rope drag on the ground or become entangled in any part of your body.

3 Pass the strap from the right hand into the left.

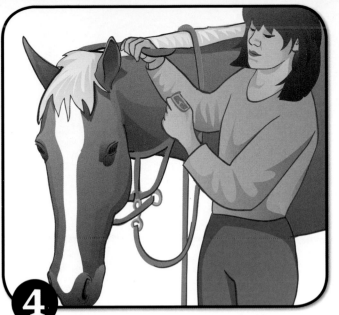

4 Hold the buckle in the left hand and the strap in the right.

5 Lower the nose piece of the halter down past the horse's nose. Slip the halter around the nose.

6 Pull up the strap in the right hand and work the halter up the left side of the horse's face. Buckle the halter just below the left ear.

SAFETY TIP
Move slowly around your horse's head, because sudden movements may frighten him.

TAKING CHARGE

HORSES LOOK FOR A LEADER, AND IT IS YOUR JOB TO BE IN CHARGE. They are social animals that exist in a hierarchical herd. You must be above your horse in the hierarchy, and you accomplish this by making him respect you during ground work.

ESTABLISHING PERSONAL SPACE

Horses can be courteous and must have manners, just as people do. Here are some guidelines.

- **YOUR HORSE MUST NOT RUB HIS HEAD ON YOU** as if you are a rubbing post. This gesture is a sign of disrespect; it is something he would do to a horse that is lower in the herd hierarchy.

- **YOUR HORSE MUST STAY IN HIS OWN SPACE** in the area to your right.

- **YOUR HORSE MUST BE OBEDIENT,** stopping beside you as well as following where you lead. If he does not stop next to you when you stop, pull abruptly on the lead rope and make him stop sharply, then turn toward him and use the lead rope to make him back up.

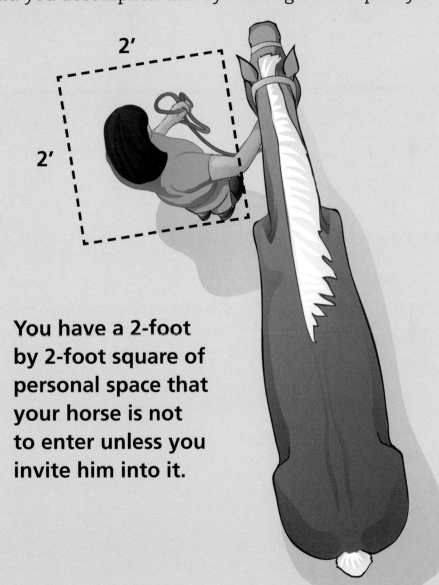

You have a 2-foot by 2-foot square of personal space that your horse is not to enter unless you invite him into it.

Safe Positions for Leading

Tools you will need: *halter, lead rope*

IN ORDER TO MOVE A HORSE FROM ONE PLACE TO ANOTHER, you must be able to steer him and stop him by means of a halter and a lead rope. The halter controls the horse's head and allows you to steer him in the direction you want to go.

HELPFUL TIPS

- Make sure that your lead rope is at least 10 feet long.

- Always lead horses from the left side.

- Do not wrap the rope around any part of your hand or body.

- When leading, always keep a safe distance of at least 15 feet from other horses.

- Do not fall back behind the horse's shoulder when leading.

- Do not let the horse walk directly behind you, because if he startles, you will be in his escape path.

CORRECT POSITION

Stand on the left side of the horse, with your right hand holding the lead rope 4–5 inches down from the halter, and the excess rope in a figure 8 in your left hand.

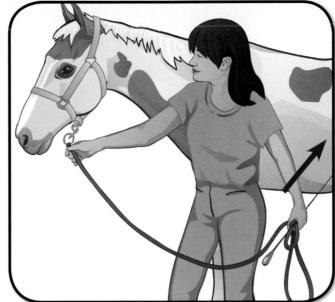

WALKING. Walk on the left side of the horse, staying between his shoulder and his ear. Look ahead and not at your horse. Say "cluck" to the horse with a slight forward tug on the lead rope.

STEPPING OUT. If your horse lags behind you, use a long dressage whip in your left hand and reach behind you to touch his hip with it.

STOPPING. To halt the horse, stop and pull on the lead rope with your right hand toward the middle of his chest. Say "whoa" calmly but firmly. Reward him when he stops by relaxing your pressure on the rope.

HELPFUL TIP

If the horse does not immediately walk forward, pull the lead rope gently forward and toward one side asking him to step forward toward you. This will shift his balance to one side and encourage him to walk with you.

HELPFUL TIP

Do not pull the lead rope toward you when stopping the horse as it will bring him into your personal space.

Backing Up Your Horse

Tools you will need: *halter, lead rope*

LOOK THE HORSE STRAIGHT IN THE EYE as you ask him to back up. This dominant posture is necessary to gain his respect. If he is being obstinate about backing up, follow the steps on the page below.

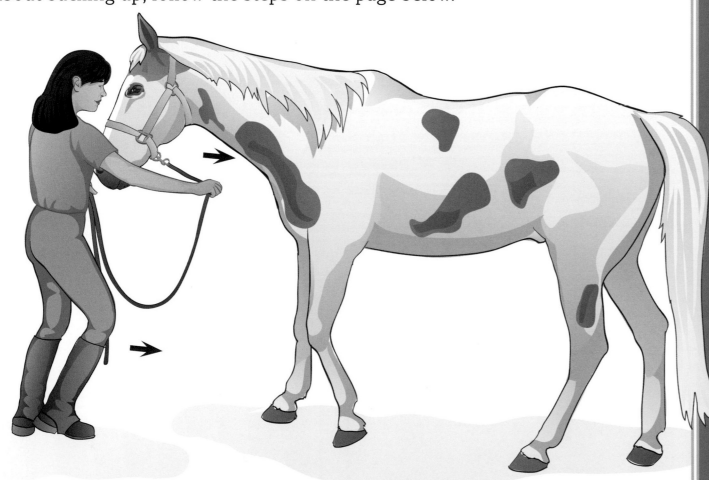

FACE YOUR HORSE while standing at the front edge of the safe zone (see page 4). Pull the rope away from you and toward his chest, firmly say "back," and step toward his head.

SAFETY TIPS FOR LEADING

- When walking with someone who is leading a horse, always walk on the same side of the horse as the person who is handling the horse.

- Keep the horse that you are leading a safe distance from other horses.

- When passing another person leading a horse, always pass with your left shoulder to the other person's left shoulder.

- If a horse crowds your personal space while you are leading him, push your elbow and forearm against his neck to move him away.

BACKING UP AN OBSTINATE HORSE

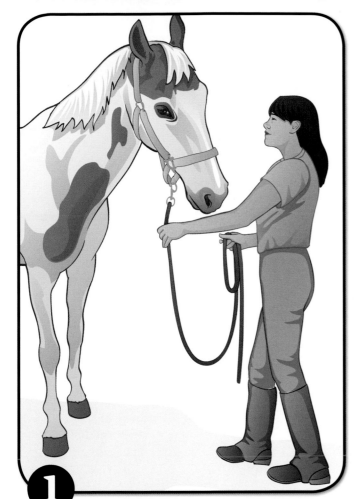

1 Hold the lead rope with the left hand close to the horse's head and the excess rope in the right.

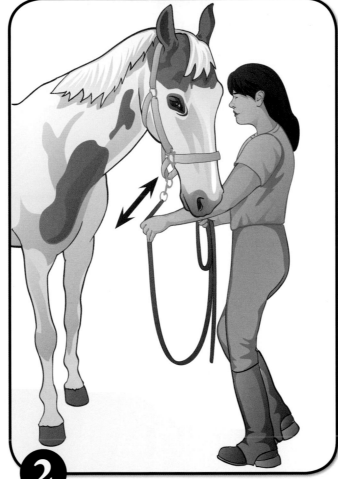

2 Release the pressure to give slack in the rope, then snap it with your left hand toward the horse's chest. The halter should thump down on his nose and make him want to move away.

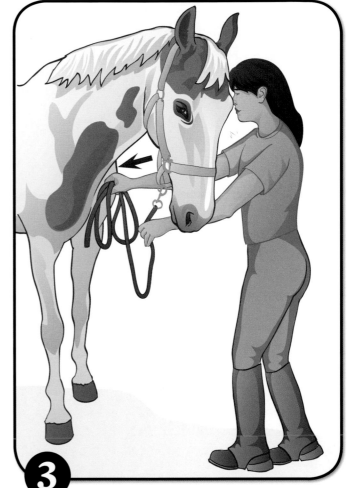

3 With your right hand holding the excess rope, push on his chest to make sure he gets the message to move back.

Leading an Excited Horse

Tools you will need: *halter, lead rope*

WHEN A HORSE BECOMES EXCITED, maintain control and leadership through your body language. Use an authoritative body posture and look steadily in his left eye. Facing the horse's left shoulder, hold the lead rope in your left hand, 4–5 inches down from the halter, with the excess in your right hand.

1

KEEPING HIM OUT OF YOUR PERSONAL SPACE. Push your right arm into the place where his shoulder blade meets his neck.

2

CALMING HIM DOWN. Push on the horse's shoulder with your right hand, while your left hand draws him in a small circle around you.

3

STOPPING HIM FROM CROSSING IN FRONT. Use your left hand in his line of sight, level with his eye, to move his head and keep it in his own space to your right.

Emergency Lip Rope

THE EMERGENCY LIP ROPE is one way to control an overly excited horse.

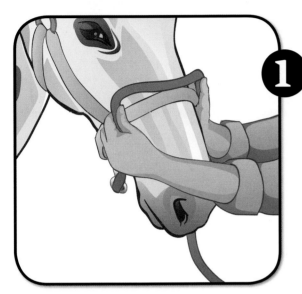

1 With the lead rope attached to the bottom ring of the halter, loop the rope over the nose piece of the halter from right to left.

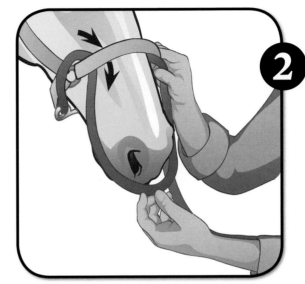

2 Slide the rope under the nose piece so it hangs down in front of the horse's nose.

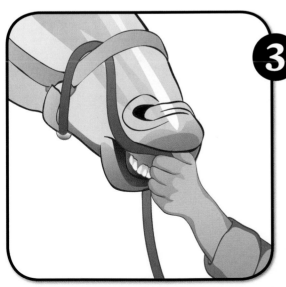

3 Put this loop of lead rope under the horse's upper lip.

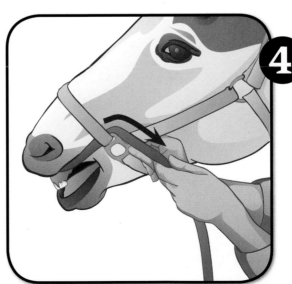

4 Cinch the rope down tightly. Now lead the horse directly back to his stall where you can remove the lip rope and allow him to quiet down.

Using a Chain

A HALTER AND LEAD ROPE MAY NOT BE ENOUGH to control a very spirited or excited horse. A chain can be an extremely effective tool for establishing respect and control; however, it must be used with caution because of the amount of force it applies to the animal's face. Be especially careful if you have never used a chain on your horse.

AROUND THE NOSE

This configuration is most effective because it puts pressure on the horse's nose and will always loosen when pressure is released.

Tools you will need: *halter, lead rope, shank chain or lead shank*

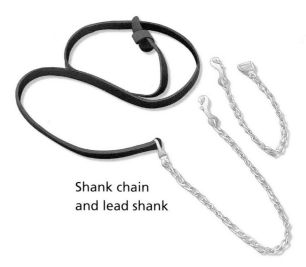

Shank chain and lead shank

① **Run the snap through the ring on the left side of the halter nearest to the horse's mouth, from the outside to the inside.**

② **Bring the chain over the nose piece of the halter.**

CAUTION

Never tie a horse that has a chain on the halter. The horse may become agitated, and if he resists against the chain, he may injure himself.

3 Run the snap through the ring on the right side of the halter closest to the horse's mouth, from the inside to the outside.

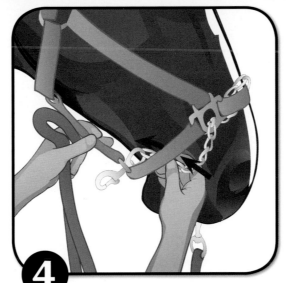

4 Run the snap down through the bottom ring of the halter from the inside to the outside.

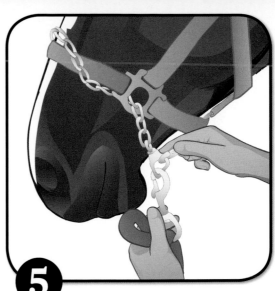

5 Snap the chain back to itself at the same place that the lead rope is attached.

HELPFUL TIPS

- **USE A 30–32-INCH CHAIN** so it will be long enough to reach around the nose.

- **SNAP YOUR LEAD ROPE TO THE END OF THE CHAIN** before putting it on the halter, if your shank chain is not attached.

- **THE BEST TECHNIQUE FOR USING A CHAIN** is to lead with a gentle hand that gives the horse slack on the lead rope and applies pressure to the chain on the horse's head if he begins to act up.

- **MAKE SURE TO HOLD THE LEAD ROPE AND NOT THE CHAIN** itself when leading.

More Ways to Use a Chain

Tools you will need: *halter, lead rope, shank chain or lead shank*

HERE ARE SOME OTHER WAYS TO USE A CHAIN with an overexcited horse. Each method has a potential problem, so be extra alert and aware. Never tie a horse with a chain on his halter.

UNDER THE JAW

This configuration applies pressure to the sensitive area under the jaw, so it can be very effective, but it can also encourage the horse to rear if too much pressure is applied.

❶ Run the snap through the ring on the left side of the halter nearest to the horse's mouth, from the outside to the inside.

❷ Run the chain under the jaw, then through the ring on the right side of the halter, from the inside to the outside.

❸ Depending on the length of the chain, it can be snapped to the ring closest to the horse's mouth on the right side or to the ring closest to the horse's right eye (shown).

OVER THE NOSE

This configuration applies pressure to the bridge of the nose, but it can turn the halter around, moving the cheek piece of the halter into the horse's right eye.

1 Run the snap through the ring on the left side of the halter nearest to the horse's mouth from the outside to the inside.

2 Run the chain over the nose, then through the ring on the right side of the halter closest to the horse's mouth from the inside to the outside.

3 Depending on the length of the chain, you can snap it either to the ring closest to the horse's mouth on the right side or to the ring closest to the horse's right eye (shown).

HELPFUL TIP

If your horse runs backward and away from the chain, do not pull on the lead rope to stop him. Follow him with a slack in the rope, and he will stop sooner.

Leading & Turning Loose in the Stall

Tools you will need: *halter, lead rope*

BE CAREFUL WHEN REMOVING THE HALTER INSIDE THE STALL.
Many horses are in a rush to be turned loose and will try to overpower the handler. In this situation, firm and methodical handling will make a horse polite and safe to release. Never turn your back on a loose horse in a stall.

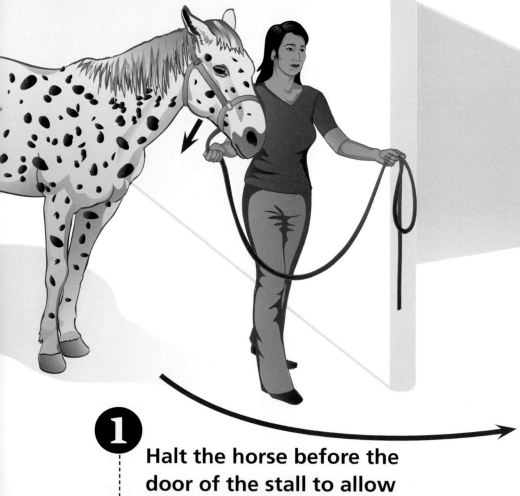

❶ Halt the horse before the door of the stall to allow you to walk in first.

❷ Lead the horse all the way into the stall. Then turn him toward you so that you are between him and the

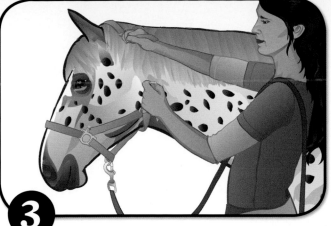

3 Put the lead rope over your shoulder. Stand on his left and unbuckle the halter, keeping the strap in your right hand and the buckle in your left.

4 Hold the buckle and strap while slipping the halter off the horse's nose.

5 Let go of the strap with your right hand so that the halter falls away from the horse's neck but the buckle remains in your left hand.

HELPFUL TIPS

If your horse tries to rush away when you unbuckle the halter, keep the buckle in your left hand and the strap in your right. Hold the halter tightly around his neck near his head until he relaxes. Then release the halter.

6 Step away from the horse, still facing him, and back out the door. Close and latch the door.

GRAZING YOUR HORSE SAFELY

GRAZING CAN BE A RELAXING AND REWARDING EXPERIENCE for both you and your horse as long as some safety precautions are taken. Keep the following in mind while grazing your horse.

1 **KEEP THE LEAD ROPE OFF THE GROUND** so that the horse cannot step on it and spook himself.

2 **IF THE HORSE DOES GET SPOOKED, DON'T PULL ON THE LEAD ROPE.** Hold the loose end of the lead rope gently and follow your horse as he runs backward. Reassure him that he is safe and quietly lead him back to a good grazing spot.

3 **ALWAYS STAY NEAR YOUR HORSE'S LEFT SHOULDER,** taking care not to let the horse move into a position where you are at his flank or behind him.

4 **NEVER SIT, KNEEL, OR LIE DOWN WHILE YOUR HORSE IS GRAZING,** so that you can quickly adjust your position if he moves.

5 **STAY ALERT.** Sudden movements or loud noises like a passing vehicle may startle him.

6 **IF THE HORSE BECOMES AGITATED,** lead him in a small circle around you, switching the lead rope to your left hand and placing your free hand on his left shoulder to keep him at a distance.

7 **APPLY THE EMERGENCY LIP ROPE** (see page 19) if the horse remains agitated after being led in a circle. Then return him safely back to the barn.

Turning Loose in a Corral

Tools you will need: *halter, lead rope*

MOST HORSES ARE VERY EXCITED with the prospect of being allowed to run free in a turnout arena. To ensure the safety of you and your horse, you must establish control before you turn the horse loose.

1 Halt the horse before the gate of the corral to allow yourself to walk in first.

2 Lead the horse into the corral, then turn around and latch the gate behind you.

3 Walk to the middle of the corral.

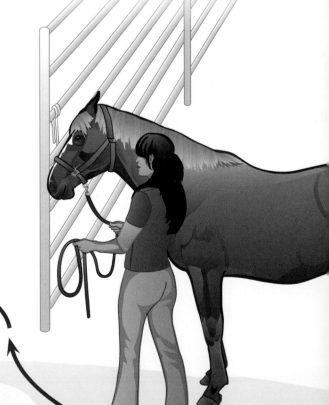

4 Put your lead rope over your shoulder. Stand on the horse's left side, facing his left ear. Unbuckle the halter, keeping the strap in your right hand and the buckle in your left.

5 Hold the buckle and the strap while slipping the halter off the horse's nose. The halter will still be around the horse's neck.

6 Release the strap with your right hand as you back away from the horse. The halter falls away from the horse's neck but the buckle remains in your left hand.

DEALING WITH A LOOSE HORSE

- If you know that your horse tries to break away when you unbuckle the halter, put the rope over his head and grasp it under his neck before you unbuckle the halter. Keep the rope in your right hand and hold it close to his head to control him.

- Always back away from a loose horse. If the horse comes in your direction, swing the halter and lead rope toward him as though you might strike him.

- Never approach the rear end of a horse that is loose.

Tying Your Horse

Tools you will need: *halter, lead rope, and tie rail*

WHEN SINGLE-TYING YOUR HORSE WITH A ROPE, leave 2½–3 feet of rope between his head and the tie rail. If the rope is any shorter, it may cause him to become upset and pull back. If the rope is too long, the horse may step over it or catch it behind his ears. Always tie your horse at a level at or above your horse's chest.

QUICK-RELEASE KNOT FOR THE SINGLE TIE

1 Double the rope approximately two and a half feet away from the horse's head and lay it over the tie rail.

TYING TRICKS

Never walk between the tie rail and a tied horse; always pass in front of the rail or safely behind the horse.

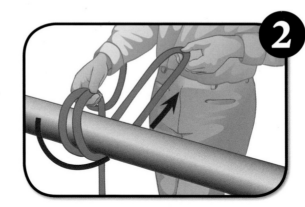

2 Pull the loop of rope under the tie rail, keeping one hand on the doubled rope above the tie rail and the other hand on the loop.

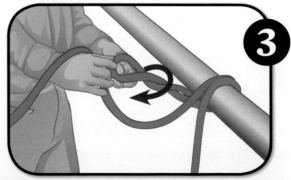

3 Pull the loop up above the tie rail and twist it twice.

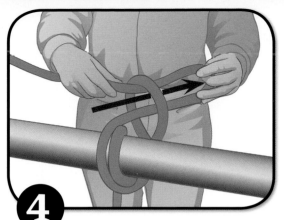

4 Make a small loop out of the rope attached to the horse's halter and push it through the loop created by the twist.

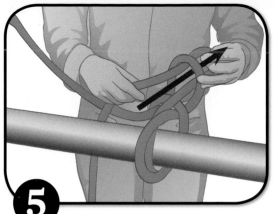

5 Pick up the loose end and make a loop. Push the loop through the last loop created by the piece of rope attached to the horse's halter.

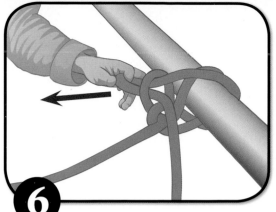

6 Pull the last loop you made back toward the horse's head to tighten the knot.

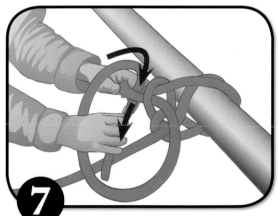

7 Loosely push the loose end of the rope through the last loop to lock the knot.

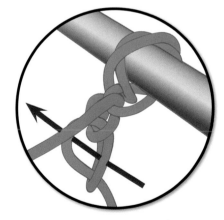

QUICK RELEASE.
To release, move the loose end back through the last loop and pull firmly.

MANAGING A SPOOKED HORSE

Follow these instructions if a single-tied horse is pulling back.

1 Get a safe distance behind the horse and try to chase him forward.

2 Once he has relieved the tension on the rope, unsnap the tied lead from the halter.

3 Snap another lead rope to the halter and hold the horse until he calms down.

4 Wrap the lead rope around the tie rail several times to simulate tying. Don't walk away.

Securing Your Horse

Tools you will need: *halter, lead rope, and cross-ties*

CROSS-TIES

A horse must become accustomed to cross-ties if he has only been single-tied. Make sure that the cross-ties are long enough to allow the horse to move his head, but not so long that he can turn around.

1 Lead the horse past the cross-ties, closer to the right wall than to the left wall.

2 Turn and face the horse. Walk backward toward the cross-tie on his left side, drawing him toward you.

CROSS-TYING TRICKS

- Keep your fingers out of any tight spaces, like the rings of the halter, where they could become caught if the horse is startled.

- Never tie a horse to the bit in his mouth. Always attach a lead rope or cross-ties to the halter.

- If you need to tie your horse after you have put his bridle on, put the halter over the top of the bridle and tie the cross-ties to the halter.

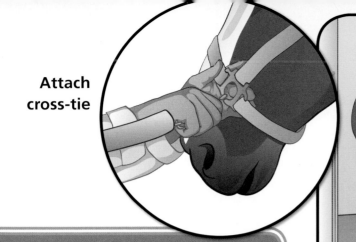

Attach cross-tie

MANAGING A SPOOKED HORSE

Follow these instructions if a horse is pulling in cross-ties. Make sure that you always have room to move out of the way.

❶ Move away from the cross-ties until the horse has finished struggling.

❷ When the horse is quiet, be very careful as you approach because the horse may begin thrashing again at any time.

❸ Reach in slowly from the side and attach a lead rope to the halter. Gently unsnap the cross-ties one side at a time.

❸ You are now standing at the horse's left shoulder toward his head. Hold the horse's halter in your left hand. Use your right hand to attach the cross-tie to the ring on the left side of the halter closest to the horse's mouth (see inset above).

❹ Step across the front of the horse to his right side. Hold the halter in your right hand. Use your left hand to attach the cross-tie to the ring closest to the horse's mouth on the right side of the halter.

SAFETY TIPS WHILE GROOMING

GROOMING YOUR HORSE IS A BONDING EXPERIENCE FOR BOTH OF YOU. Nevertheless, your horse can unintentionally injure you if you are working in an unsafe area. Stay out of the danger zones directly in front of and behind the horse. You can reach into these spaces, but keep your head and body out of the way. Follow these tips.

1 **IT IS SAFE TO WORK AT THE SIDES OF THE HORSE** and at an angle to the shoulder and the hip (see illustration on page 4).

2 **IF YOU DROP A TOOL UNDERNEATH** the horse, do not reach below to retrieve it. Instead, gently kick the tool clear of his legs, then bend down to pick it up.

3 **WHEN WORKING ON YOUR HORSE'S LEGS,** place your free hand just above where you are working. This way, you will feel any motion coming and can push away if there is a problem.

4 **NEVER KNEEL OR SIT BESIDE YOUR HORSE.** Keep at least one foot on the ground so that you can be ready to move out of the way at any time.

5 **NEVER BEND DOWN** behind a horse's hind legs.

6 **KEEP ALL GROOMING TOOLS, TRAYS, AND BUCKETS** on the same side of the horse that you are working on. Make sure to take them with you when you change sides.

Grooming Tools

A FEW BASIC GROOMING TOOLS, shown on the following pages, are standard equipment for maintaining a healthy coat. When using these grooming tools be sure to stay away from the sensitive areas of the horse, outlined below.

Tools you will need: *curry comb, stiff-bristled brush, soft-bristled brush*

BENEFITS OF GROOMING

Daily grooming does good things for your horse. For example:

- **REMOVES DIRT,** sweat, dead skin cells, and hair.

- **FACILITATES SHEDDING** and adds a glossy sheen to his coat.

- **ALLOWS A THOROUGH INSPECTION** of your horse, which may help you to detect any health problems.

- **ACCUSTOMS YOUR HORSE TO BEING HANDLED.**

- **PROVIDES A GOOD BONDING EXPERIENCE** for you and your horse.

A horse's most sensitive areas are highlighted.

face

flank area

armpit area

CURRY COMB

CURRYING BRINGS UP NATURAL OIL TO MAKE THE COAT SHINE.

Use the curry comb on the fleshy parts of the body, working in a circular motion with enough pressure to massage the roots of the hairs as shown below. The curry comb can be made of rubber or metal and is used to loosen dirt and old hair from the horse's coat.

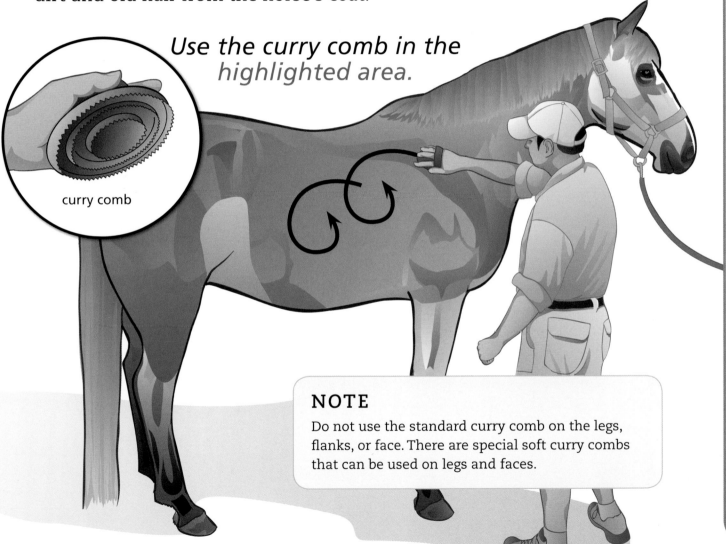

Use the curry comb in the highlighted area.

curry comb

NOTE

Do not use the standard curry comb on the legs, flanks, or face. There are special soft curry combs that can be used on legs and faces.

HOT-TOWEL TECHNIQUE

After a thorough currying, this hot towel technique will remove loosened dust and dirt from the coat. It also opens the pores to help the natural oils enhance the coat and make it shine. You will need a small bucket of warm water with a hand towel in it. Add ¼ cup of liquid hair polish to the water.

❶ Wring out the towel and rub one quarter of the horse vigorously with it, picking up any dirt that was loosened while currying.

❷ Rinse out the towel and wring it out again. Repeat Step 1, working on one quarter of the horse at a time until the entire coat has been cleaned.

❸ Let the horse dry. Then use the stiff and soft brushes to smooth and polish the coat.

STIFF BRUSH

NEXT USE THE STIFF BRUSH, starting just behind the horse's ears and working your way back to his tail. Brush with a flicking motion that smoothes down the hairs. You can use the stiff brush carefully on the horse's legs. This brush's stiff bristles remove the dirt and hair brought up by the curry comb.

DETANGLING THE TAIL

- Generously spray the tail with hair detangler before beginning to untangle the tail. A comb or hair brush is good for getting the tangles out of the mane, tail, and forelock.

- Always start from the bottom of the tail and comb out that section, then move up to the next section and so on.

- For a badly matted tail, hold the tail in one hand and pick out a few individual hairs at a time. Repeat with just a few hairs at a time until you have untangled the entire tail.

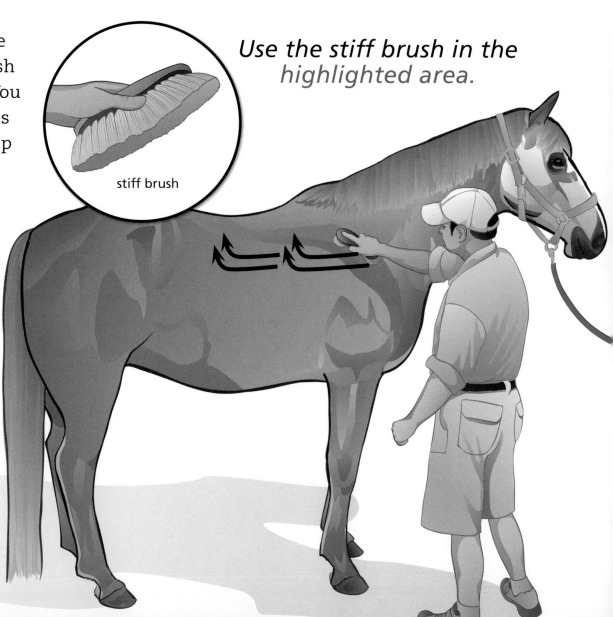

stiff brush

Use the stiff brush in the highlighted area.

SOFT BRUSH

NOW USE THE SOFT BRUSH with long sweeping strokes to further smooth and polish the hair and remove any final dust. You can also use the soft brush on the face, using caution not to poke it in the horse's eyes or nose.

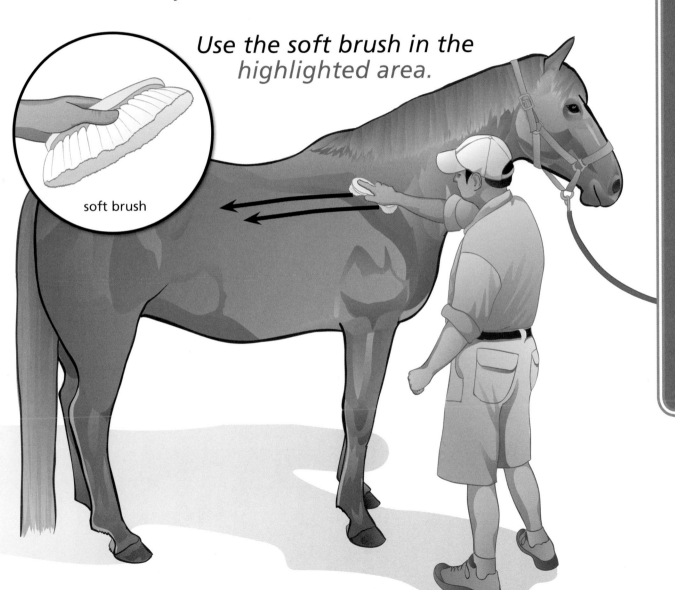

Use the soft brush in the highlighted area.

soft brush

Caring for Hooves

KEEPING YOUR HORSE'S FEET IN GOOD CONDITION is of paramount importance to his health and soundness. To do this, you must know how to identify parts of the hoof, how to pick up the horse's front and hind legs, and how to pick out his hooves.

UNDERSTANDING BASIC FOOT ANATOMY

Familiarize yourself with the parts of the hoof and the lower leg, which contains important bones that are vitally connected to the hoof.

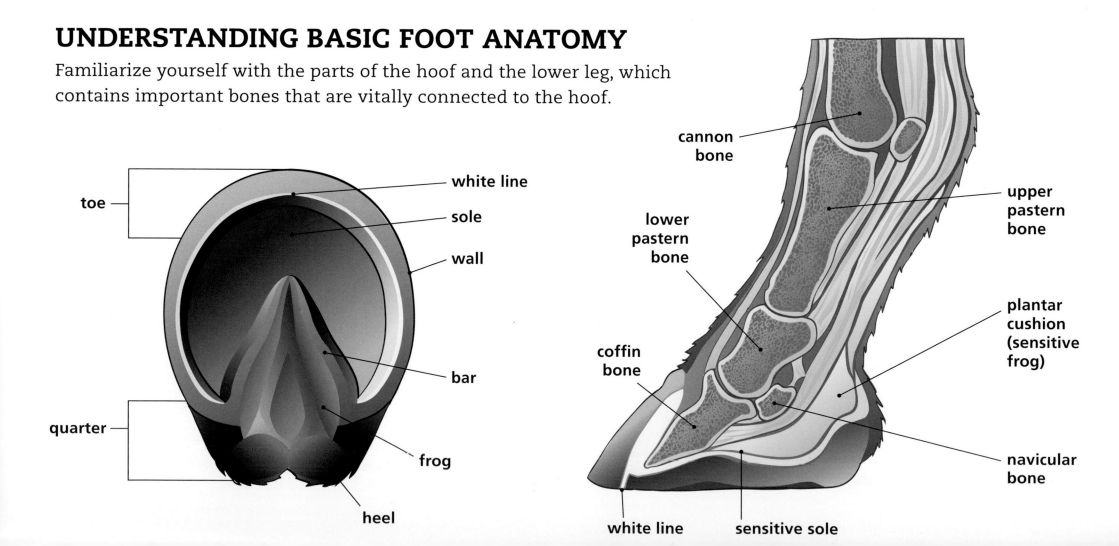

toe

white line

sole

wall

bar

quarter

frog

heel

cannon bone

lower pastern bone

coffin bone

upper pastern bone

plantar cushion (sensitive frog)

navicular bone

white line

sensitive sole

PICKING UP A FRONT LEG

1 Stand with your hip next to the horse's shoulder, facing the rear of the horse. Slide your hand down the back of the horse's leg, squeezing the tendon between the horse's knee and his fetlock.

2 As the horse takes the weight off that foot, carefully lift it off the ground and hold it, sole up, by the pastern or hoof.

PICKING UP A HIND LEG

1 Stand with your hip near the horse's hip facing the rear of the horse. Slide your hand down the back of the horse's leg, squeezing the tendon between the horse's hock and his fetlock or hoof.

2 As the horse takes the weight off that foot, carefully lift it off the ground and hold it, sole up, by the pastern.

PICKING OUT A HOOF

The hoof pick is used to dig out any dirt, shavings, or rocks that might be in the horse's foot. Use the hoof pick with enough pressure to dislodge anything in the hoof. Pick out your horse's feet before and after a ride.

Tools you will need: *hoof pick, hoof oil brush, hoof sealer*

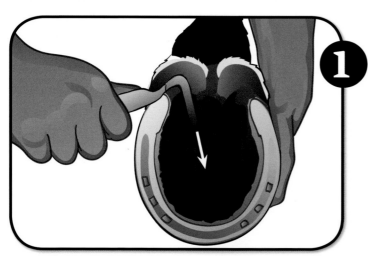

1 Holding the foot up, draw the hoof pick with firm pressure from the corner of the frog to the toe, loosening and removing any dirt or debris.

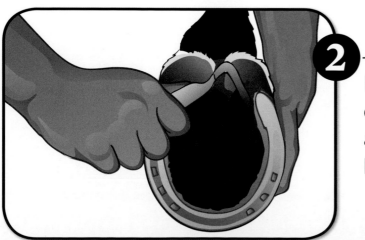

2 Make sure to dig into the crease between the frog and the bars, especially back toward the heel.

HELPFUL POINTS ON PICKING

- Make sure to use enough force to loosen debris. The horse's foot is very hard and can withstand the scraping of a hoof pick.

- Never pick a horse's feet while standing behind him.

- Keep your face and head away from the feet or from the line a hoof might travel if the horse moves its leg.

- Watch to make sure that your own feet are clear of the horse's when you let go of the foot after picking it out.

- Horses' hooves should be trimmed or re-shod every six weeks.

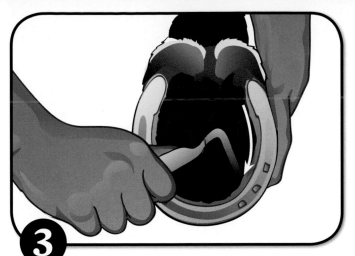

3 Run the hoof pick under the edge of the shoe, then use a stiff plastic brush to brush out any remaining dirt.

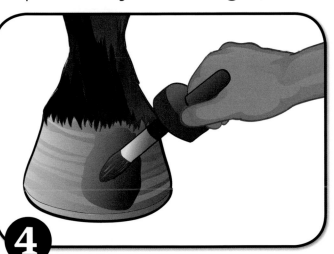

4 Use the hoof oil brush to paint the bottom of the foot. Then paint the hoof wall from the coronary band to the ground.

REFERENCE GUIDE TO COMMON HOOF PROBLEMS

Thrush	A foul-smelling anaerobic bacterial infection of the frog. Seldom causes lameness.
Sand Cracks	Vertical cracks that can be superficial or deep; may require corrective trimming/shoing and use of a hoof preparation such as hoof oil. May not cause lameness.
Navicular Disease	The result of bone and/or soft tissue changes in the horse's heel, causing pain. Must be diagnosed by a veterinarian and can often be addressed with corrective shoeing.
Laminitis	An inflammation of the sensitive lamina connecting the hoof wall to the coffin bone, resulting in extreme pain and lameness. Requires immediate attention by a veterinarian.
Abscess	A small pus-filled infection that occurs in the sole or white line of the foot, causing pain and lameness. Usually relieved by a veterinarian or farrier by allowing the infection to drain.
Contracted Heels	Term for when the width of the heel is less than $2/3$ of the width of the widest part of the hoof. May or may not cause lameness, but requires corrective shoeing.

Basic Clipping

CLIPPING THE HORSE CREATES A MANICURED LOOK. Most horses in shows are clipped. Some horses receive a whole body clip in the winter months when their hair is long, but this is a major job. The steps illustrated below show the typical clipping with small hand-held clippers done for general maintenance.

Some horses are fearful of clippers, so introduce them slowly and carefully the first time. When you first present the clippers, hold them with your finger extended beyond the blades to touch the horse and let him know what you are doing.

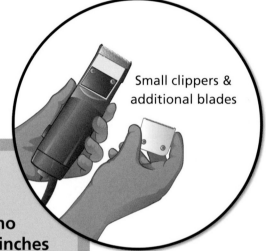

Small clippers & additional blades

CLIPPING TIPS

- Use caution when moving clippers around the horse's ears, as most horses are wary of clippers in this area and can react violently.

- Use a light touch when removing the whiskers that have face hair around them so as not to remove any hair other than the whiskers.

- Some people prefer to leave the whiskers long so that the horse can use them as natural sensors.

- Use a small clip-on comb (purchased at any beauty supply store) to make the clipping on the jaw and fetlocks even more attractive.

- Clip the bridle path to a length that is no longer than the horse's ear (about 2–3 inches for most breeds), although certain breeds require a longer bridle path or none at all.

- You can trim the ergots off close to the skin with a pair of garden shears.

- Make sure that the horse does not step on or chew the electrical cord while you are clipping.

- Keep checking the clipper blades to make sure they are not getting too hot, and use clipper lube frequently to keep the blades clean and running smoothly.

 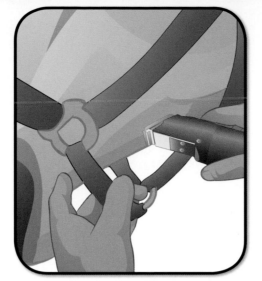 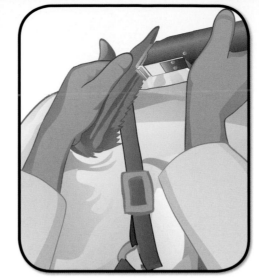 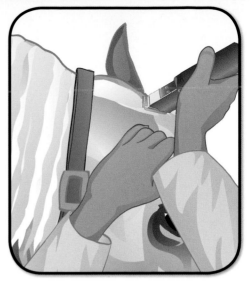

MUZZLE. Keep clipper blades parallel to the horse's skin, but make sure they rest on the skin. Remove all of the long hairs on the muzzle with a size 40 clipper blade.

JAW. With a size 10 blade, shorten the hair under the horse's jaw, starting at the throat and moving down to the base of the chin.

EAR. Hold the horse's ear closed with one hand so that the edges touch. Use a size 10 blade to remove the hair that is sticking out and the hair along the edge of the ears.

BRIDLE PATH. Press a size 40 blade against the skin at the poll between the ears and move down the mane about 2 inches. Then clip against the hair back toward the poll.

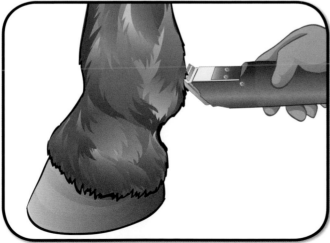

LEG. With a size 10 blade, start at the back of the pastern and clip upwards around the ergot to the edges of the fetlock. Use a light touch as you move up the back of the leg along the flexor tendon to the knee.

Bathing a Horse

Tools you will need: *bucket, sponge, shampoo, scrubbing mitt, sweat scraper*

BATHING YOUR HORSE WITH SHAMPOO should be done before a show if the weather allows, but using soap too often strips some of the natural oils from his coat. Showering your horse with water can be a daily affair if it's warm enough outside. Fill a bucket three-quarters of the way full. Add about ½ cup of concentrated shampoo or 1–1½ cups of regular shampoo. Mix with the sponge to form a lather.

SOAPING UP

1

START BEHIND THE EARS NEAR THE MANE. With sponge in one hand and scrubbing mitt in the other, squeeze the soapy water out as you drag the sponge down the neck, in the direction the hair grows.

2

WORK THE SUDS into the mane and hair on the neck with the scrubbing mitt. Repeat down the length of the body and legs, soaping one section at a time, then scrubbing with the mitt.

3

HOLD THE BUCKET near the side of the tail and carefully dip the long part of the tail in the soapy water.

CAUTION
Make sure to scrub each of the legs gently as the skin is sensitive.

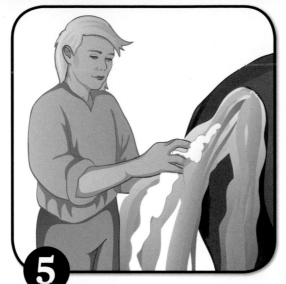

4 SET THE BUCKET ASIDE AND USE THE SPONGE to put soapy water on the dock of the tail and along the tailbone.

5 SCRUB WITH YOUR FINGERS all along the tailbone, adding more soapy water with the sponge as needed.

6 WASH THE HEAD LAST. Wring most of the soapy water out of the sponge. Stand to the side and reach up to his head. Rub the horse's face carefully with the sponge, taking care not to get soap into his eyes or ears. Make sure to scrub the sides of the face and behind the ears.

HELPFUL TIPS

- Always keep the bucket on your side of the horse.

- For white socks or stockings, use some concentrated shampoo directly on the white hair and gently scrub until all of the dirt is gone. Then scrub with the sponge while rinsing until the pink skin is visible.

- Most horses do not like to have their head washed so use only as much soap on the face as you know you can rinse out.

- If available, use warm water to bathe and rinse your horse. If the weather is cool, put a wool cooler on him while drying off.

Bathing a Horse

RINSING

Before rinsing, sponge the remaining soapy water over the horse's back and haunches. Rinse out the bucket and sponge, then refill with clean water. You will probably need three buckets of clean water to complete the rinsing. Always move the sponge and scraper in the direction in which the hair grows.

1 **WRING OUT** most of the water in the sponge and wipe the horse's forehead. Rinse out the sponge, wring out most of the water, and sponge the sides of his face. Wring out again and sponge behind the ears. Repeat until the whole face is free of soap.

2 **NEXT TAKE THE SPONGE** without wringing it out. Start on the neck just behind the ears and squeeze the water out of the sponge as you draw it down the crest of the neck. Repeat along the top of the body.

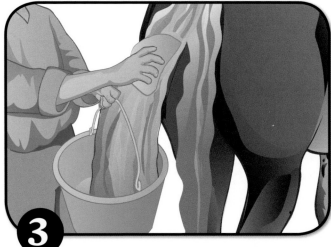

3 **DIP A SPONGE IN A SEPARATE BUCKET** half full of clean water. Place the sponge at the tail head and squeeze out the water as you draw it down the tailbone. Continue until all soap is gone. Then dip the long tail hairs in the bucket and swish until rinsed.

BUCKET & HOSE METHOD

1. Fill a bucket with water three-quarters of the way full. Add about ½ cup of concentrated shampoo or 1–1½ cup of regular shampoo.

2. Use the sponge to mix the shampoo in the water until there is a lather in the bucket.

3. First hose down the horse, avoiding his head.

4. Wash the horse as above with the bucket, soaping his body, tail, and face. Then use the bucket to rinse his face as above.

5. Use the hose to rinse the rest of the horse's body and tail. Be careful not to spray in the direction of the horse's head or ears.

6. Now use the sweat scraper on the horse's body to remove the excess water, scraping it in the direction of the hair and using your hand to push the water down and off each leg.

SCRAPING

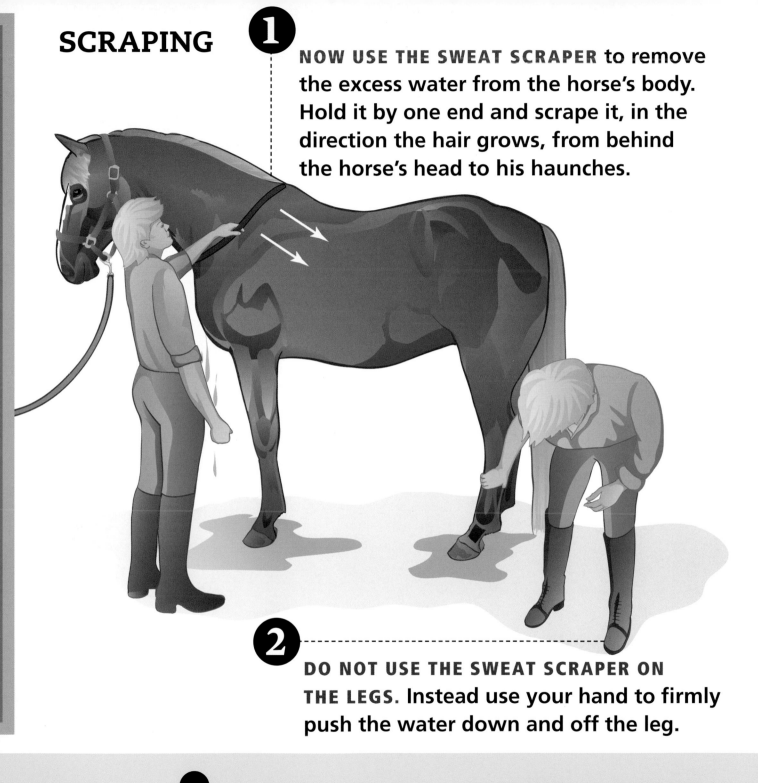

1 NOW USE THE SWEAT SCRAPER to remove the excess water from the horse's body. Hold it by one end and scrape it, in the direction the hair grows, from behind the horse's head to his haunches.

2 DO NOT USE THE SWEAT SCRAPER ON THE LEGS. Instead use your hand to firmly push the water down and off the leg.

Sheath Care

MALE HORSES have a small pouch around the head of their urethra that collects dirt, oil, and dead skin cells. This deposit forms into a hard ball called a "bean." Each year, the bean must be removed and the entire inside of the sheath cleaned. Some horses react violently to sheath cleaning and must be tranquilized by a veterinarian, so be careful.

> Tools you will need: *bucket of warm water, Latex gloves (or obstetrical gloves from your vet), roll cotton, castile soap or commercial sheath-cleaning product*

PREPARATION

Tie the horse securely. Tear off about 20 fist-sized pieces of roll cotton and put them into a bucket of warm water. Remove jewelry from the hand that you will use to clean the sheath, roll up any long sleeves, and put on a latex glove. Lather up a piece of wet cotton with some castile soap or 2–3 tablespoons of a commercial sheath-cleaning product.

SAFETY TIP

Always stand near the horse's shoulder and reach back when cleaning a sheath because he can swing his hind leg forward and sideways in a "cow kick."

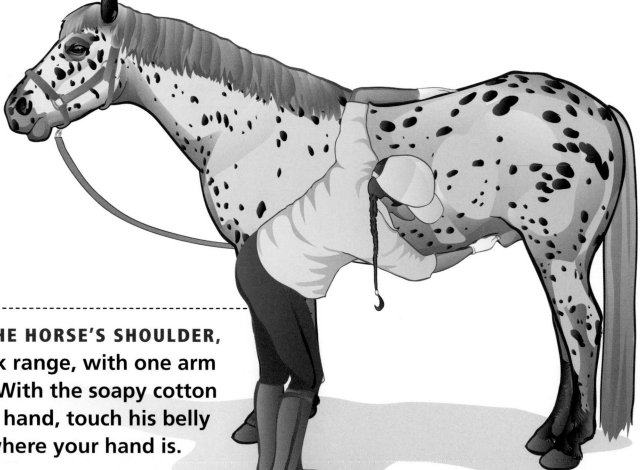

1 **STAND NEAR THE HORSE'S SHOULDER,** well out of kick range, with one arm over his back. With the soapy cotton in your gloved hand, touch his belly so he knows where your hand is.

REMOVING THE BEAN OUTSIDE THE SHEATH

If your horse drops his penis down while you are cleaning, you will not have to reach up into the sheath and can easily remove the bean from the pouch at the head of the penis.

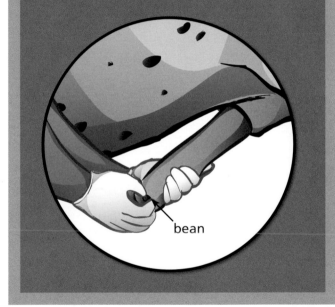

bean

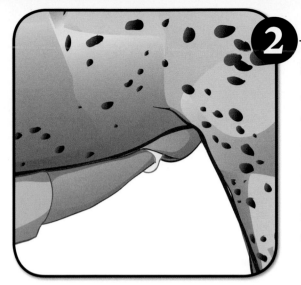

2

REACH CAREFULLY INSIDE the sheath and remove dirt and debris with the soapy cotton. When the cotton is soiled, discard it and use another piece from the bucket. Repeat the process until the cotton comes out clean and the area is well soaped.

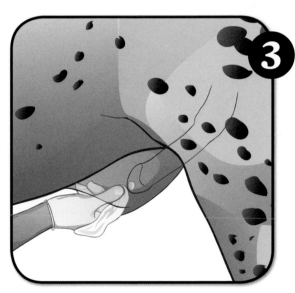

3

WITH A WELL-SOAPED PIECE OF COTTON, slide your hand further into the sheath area that holds the retracted penis. Remove the dirt and debris around the retracted penis. Feel for a hardened lump, the bean, near the tip of the penis. Gently use your soapy, gloved finger to slide out the bean and withdraw it from the sheath. Use the remaining cotton to rinse out all of the soap.

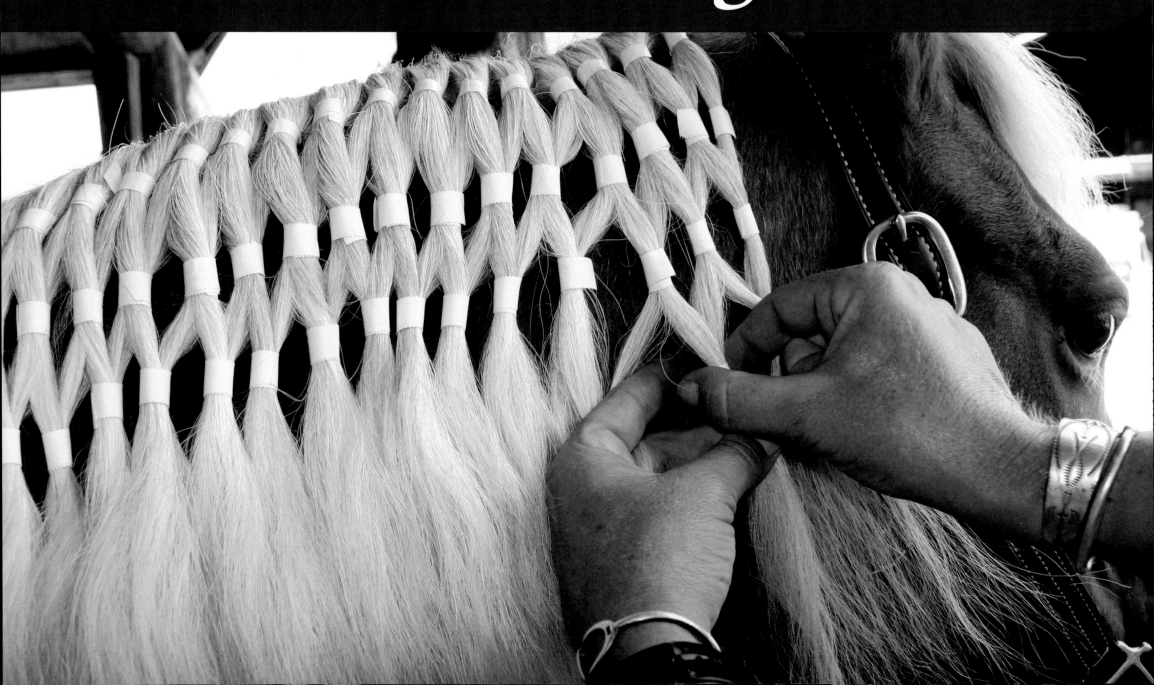

TIPS FOR SAFE & EASY BRAIDING

BRAIDING YOUR HORSE CAN ADD STYLE in the show ring and flare in photographs. Braids can show off your horse's conformation by accentuating the line of his neck or the curve of his haunches. Each discipline uses its own braiding style to present the horse to his best advantage. Here are some important tips for braiding your horse.

1 **AVOID BRAIDING YOUR HORSE AT FEEDING TIME.** He will be impatient if he thinks he is missing a meal.

2 **GIVE YOURSELF PLENTY OF TIME** to braid before a show. This way, any pre-show anxiety will not affect the quality of your braiding job.

3 **PRACTICE MAKES PERFECT.** Practice your braiding technique in a relaxed environment and time how long it takes. This will help you prepare for show day.

4 **YOUR BRAIDING NEEDS TO BE NEAT, NOT PERFECT,** unless you are in a class that is focused on grooming. The judge and spectators will observe from a distance, so it is unlikely that they will see every minute detail.

5 **USE EXTRA CAUTION** when standing on a stool for braiding. Speak softly and gently touch the horse as you carefully climb onto the stool. Reassure him again when you are getting down.

6 **WHEN MAKING THE FIRST SQUARE KNOT,** pass the string over itself twice so that it will stay snug against the hair.

7 **KEEP BRAIDS AND KNOTS AS TIGHT AS POSSIBLE** to ensure that your braids will stay in and look professional.

8 **TAKE A BREAK** and let your horse take a break too, if your fingers become tired while braiding. Chances are that he needs relief from standing still just as much as you do.

Pulling a Mane

DIFFERENT DISCIPLINES, as well as different personal preferences, call for different lengths of mane. Manes are typically shortened and thinned by pulling out the longest hairs by their roots, not by cutting the hair in a sharp line. For horses that won't allow you to pull their mane, use the alternative method on the page below.

A pulled mane

1 Grasp a small amount of mane and hold the longest hairs. Back-comb the rest of the hair toward the crest of the neck.

2 Wrap the longest remaining hairs around the comb. Hold them while you pull sharply down. Repeat this process until the mane is pulled evenly to the desired length (3-5 inches).

BANDING A WESTERN MANE

1 Start with a dampened pulled mane.

2 Use the comb to part off a portion of hair between 1 and 1½ inches wide.

3 Wind a rubber band tightly around the section roughly an inch below the crest of the neck.

4 Repeat for the length of the neck.

ALTERNATIVE METHODS

For a horse that won't let you pull his mane, follow these directions. First, either back-comb and use the blade from a large pair of clippers to cut the longest hairs, or use a razor comb to make a blunt cut. Then follow the instructions below to thin the mane, making it easier to braid.

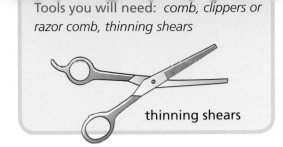

thinning shears

1 Part a small section of the mane down the crest and use one hand to hold the top portion. With the other hand make many small cuts at a 45-degree angle down the length of the hair.

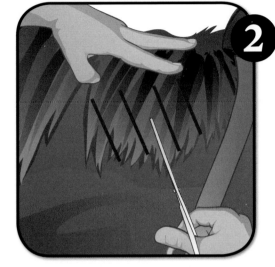

2 Place your shears at a 45-degree angle the other way and cut back down the mane. Alternate between steps 1 and 2 until the mane is evenly trimmed to a length slightly shorter than desired.

3 Flip the remaining mane over the top of the shortened portion. Hold the longest hairs in your fingers and use a pulling comb to back-comb the shorter hairs.

4 Trim the longest hairs at a 45-degree angle, alternating from right to left as in steps 1 and 2, until you reach the desired length. Repeat for the rest of the mane.

Braiding a Short Mane

THERE ARE SPECIAL BRAIDS FOR SHORT MANES. Some English disciplines require that the mane be braided for show. Some Western disciplines feature short manes banded with rubber bands to make the mane lie down neatly in the show ring.

Tools you will need: *comb, yarn cut to 2-foot lengths, scissors, hair clip, small braiding elastic bands, braid pull-through*

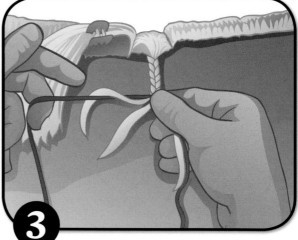

braid pull-through

ENGLISH HUNTER BRAID

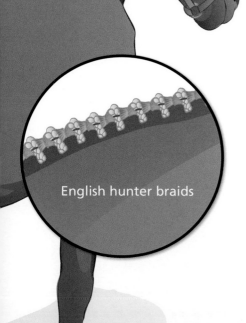

English hunter braids

1"

1 Start with a dampened pulled mane between 3½ and 4 inches long. Use the comb to part off a portion roughly 1 inch wide.

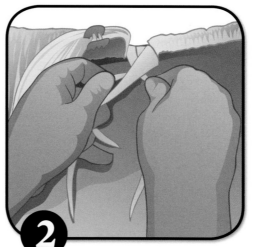

2 Use the hair clip to hold back the remainder of the mane. Start a three-strand braid.

3 Three-quarters of the way down the braid, fold the string in half and put it behind the braid. Add one side to one strand of hair and the other side to another strand.

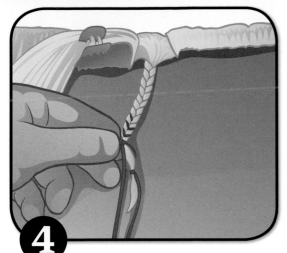

4 Continue braiding to the end of the hair. Use one end of the string to tie the braid tightly. Use the other to tie loose ends farther down.

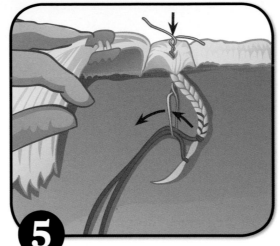

5 Push the pull-through down from the top of the braid next to the crest and feed the loose ends of the string through the loop.

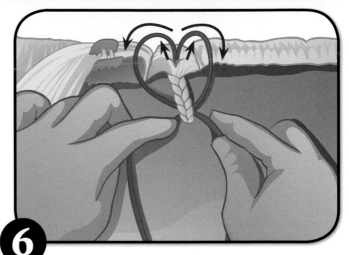

6 Pull the pull-through back through the braid along with the two strings and tie the strings in a square knot under the braid.

7 Make a tight square knot in front. Cut off the ends of the string. Repeat all steps for the entire length of the neck.

HELPFUL TIPS FOR BRAIDING OR BANDING

- You may want to stand on a stool to get a better angle.

- Put the lengths of yarn through your belt loop so they are easily accessible.

- Use a piece of tape to mark the length of the first section on the top of the comb and use this as a ruler to measure the other sections.

- If portions of your horse's mane flip over to the other side, train it to stay on one side by using large braids made of roughly 3-inch sections of hair.

- A spray bottle with water or commercial braiding spray helps keep the hair manageable.

- When taking out braids, use a seam ripper to cut the knots.

Braiding a Long Mane

DO THIS BRAID IMMEDIATELY BEFORE THE CLASS because it tends to slip out
if the horse lowers his head to eat. Keep the braid parallel to the crest of the neck as
you progress to the withers.

> Tools you will need: *comb, hair gel,
> scissors, braiding rubber bands, sponge*

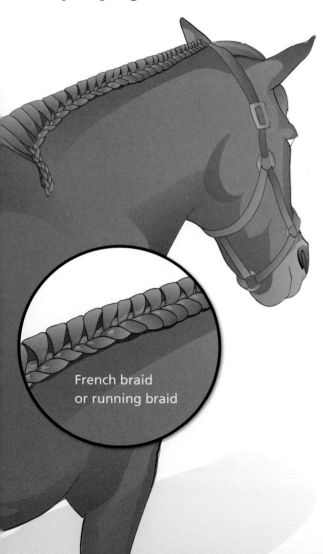

French braid
or running braid

FRENCH BRAID

1"

1 Comb out the long mane
and dampen it with a
sponge. Use your fingers
to section off a portion of
mane between 1 and 1½
inches wide.

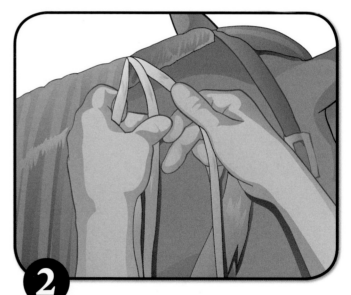

2 Start a three-strand braid
close to the crest of the
neck but pulled parallel to
the crest.

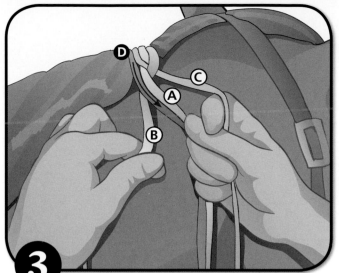

3 As you cross one strand of hair Ⓐ over the middle section Ⓑ, section off another strand Ⓓ and add it to the piece of hair Ⓐ that is crossing over the middle section.

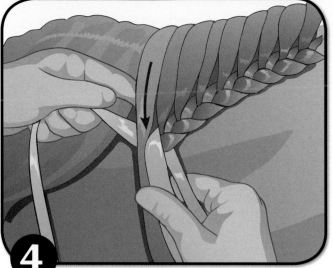

4 Add hair to the middle piece every other turn of the braid and continue braiding until you reach the withers.

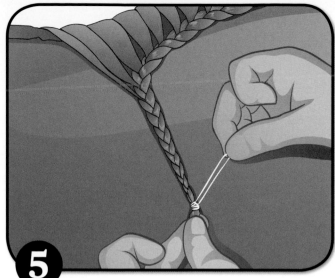

5 At the withers, do a traditional three-strand braid and secure it tightly with a rubber band or string.

6 Fold the braid up underneath itself and secure with a rubber band or tie up with string as you would a hunter braid. Use hair gel to smooth down any loose ends along the crest.

Tools you will need: *rubber bands or braiding tape, comb*

DIAMOND BRAID

This elegant braid, also known as a continental braid, looks complicated but is easy to master. It looks best with a very long mane. Make sure that the mane has been combed out completely before you start.

Finished diamond braid

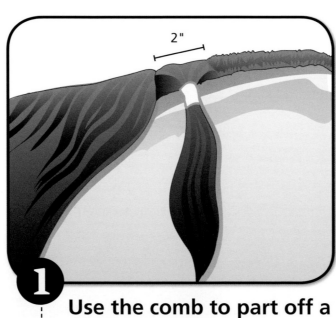

1 Use the comb to part off a portion of mane about 2 inches wide near the poll. About an inch down from the crest, wind a rubber band or a piece of braiding tape around the section.

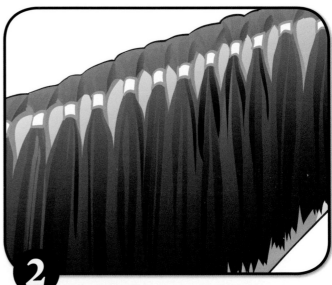

2 Working in a straight line, continue parting and rubber banding or taping uniform sections of hair until you reach the withers.

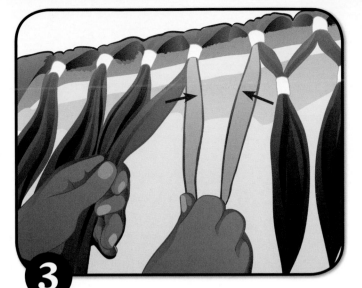

3 Return to the head. Divide each lock in half and use tape or a rubber band to connect the two nearest sections about 3 inches from the previous row.

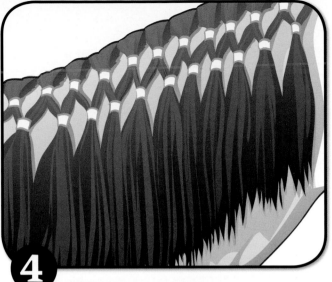

4 Continue parting sections in half and connecting those nearest to each other until you reach the withers.

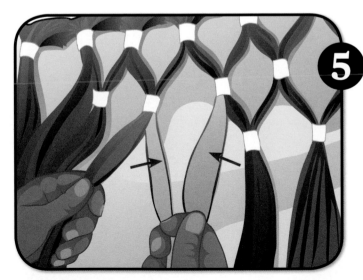

5 Repeat steps 3 and 4 until you have created many rows of diamonds (see inset on the page above).

HELPFUL TIPS

- You can use black electrician tape for a black mane and white athletic tape for a light mane; however, commercial braiding tape is available in a variety of colors.

- Twist the locks of hair together as you tape them to make a tighter connection.

- If the bands or tape are too close to the crest of the neck, the hair will stick out instead of lying flat.

- Do not make the sections too small, to make sure that the diamond pattern is visible.

- You can use a seam ripper to cut rubber bands and tape, but be careful that you do not break or cut hairs in the process.

Braiding the Tail

FRENCH BRAID

Tools you will need: *2-foot pieces of yarn, scissors, pull-through, rubber bands, comb*

THE FRENCH BRAIDED TAIL accentuates the fullness of the hindquarters. Use only on a trustworthy horse that will allow you to stand close behind him. Begin by combing out and dampening the tail hairs, especially the ones at the top and sides.

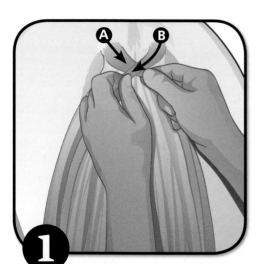

1 At the top of the tail, take a small piece of tail hair from the left **A** and one from the right **B**, then cross the right piece over the left in the middle of the tail.

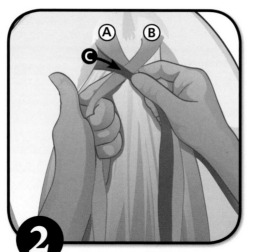

2 Hold these two pieces in your right hand. Take another small piece **C** from the left side and lay it across the top piece **B** so that three pieces are visible.

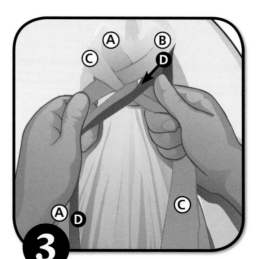

3 Twist the right piece **A** over the middle piece **C**, making it the new middle piece. Add another piece from the right side **D** to this new middle piece. Pull tightly.

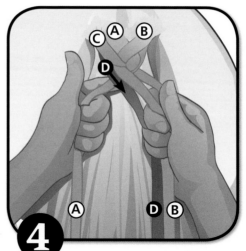

4 Twist the left piece **B** over the middle piece **A**. Add another piece **D** from the left side to this new middle piece. Continue braiding until three-quarters down the tailbone.

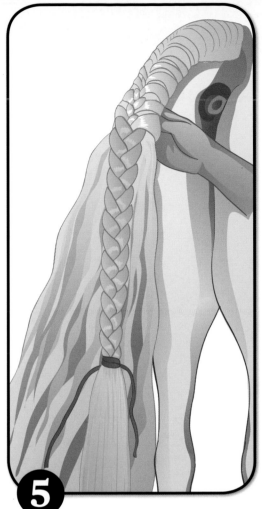

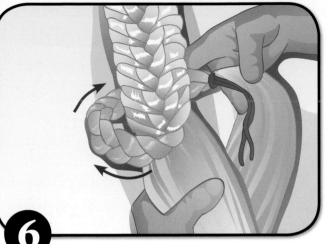

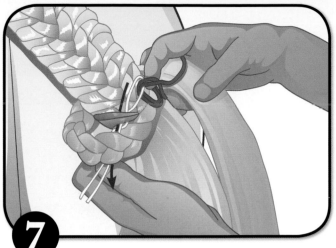

6 Turn the braid up and pass behind the tailbone.

7 Slide the pull-through under the last two left-side pieces and pass the loose yarn through it.

5 Braid a regular three-strand braid just long enough to wrap around the tailbone. Fasten with yarn.

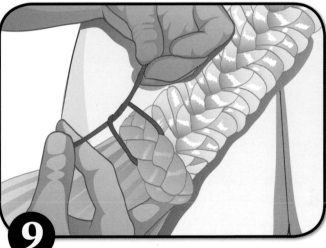

8 Pull the 3-strand braid through the last two pieces of the French braid and then under the 3-strand braid itself.

9 Tie the loose ends of the string in a square knot around the last piece of the French braid and the horizontal portion of the braid.

Braiding the Tail

Tools you will need: *rubber band, comb*

FISHTAIL BRAID

THIS IS A FUN BRAID that is not usually used for the show ring. First comb out and dampen the tail hairs.

1 At the top of the tail, take one small piece of hair from the left Ⓐ and one from the right Ⓑ. Cross the left piece over the right in the middle of the tail.

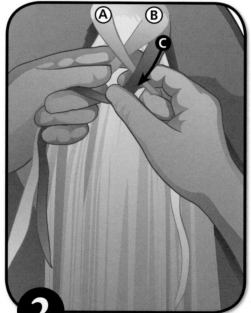

2 Hold these two pieces in your left hand and take another small piece from the right side Ⓒ. Lay it over the closest piece Ⓐ and add it to the piece on the left Ⓑ.

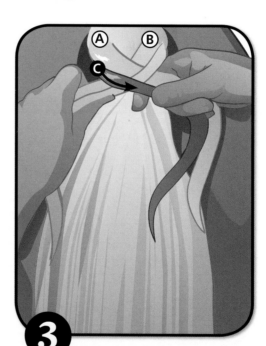

3 Hold the two pieces in your right hand. Take a small piece of hair from the left side Ⓒ, lay it over the closest piece Ⓑ, and add it to the piece on the right Ⓐ.

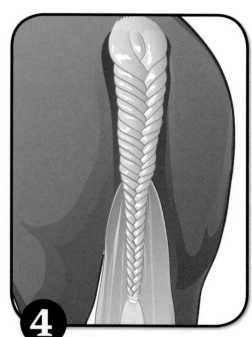

4 Continue as in steps 2 and 3 until about three-quarters of the way down tailbone. Add hair from underneath the braid to alternate sides. Secure with rubber band.

Braiding the Forelock

Tools you will need: *a 2-foot piece of string or a rubber band*

START WITH A DAMPENED, COMBED FORELOCK. Part off a small portion of hair at the start of the bridle path and divide it into three sections. If the horse does not have a bridle path, start at the top of the poll.

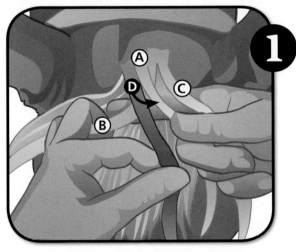

1 Add **D** to Ⓐ as you cross them over section Ⓑ, making them the new middle section.

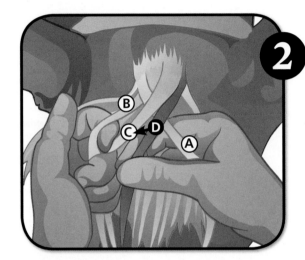

2 Add **D** to Ⓒ as you cross that section over Ⓐ, making it the new middle section.

3 Continue braiding until the end of the scalp, then do a traditional three-strand braid. Fasten with a rubber band.

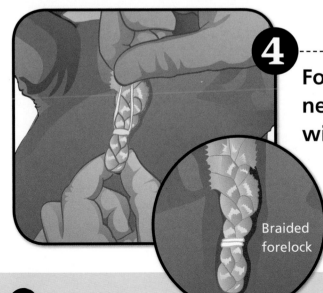

4 Fold the braid up underneath itself and secure with a rubber band.

Braided forelock

TACKING TIPS

HORSE EQUIPMENT IS TACK. The term for bridling and saddling a horse is tacking-up and removing it all is called tacking-down. Most of the tacking is done from the horse's left side.

1 **SADDLE THE HORSE** before you put on the bridle.

2 **THE GIRTH SHOULD BE TIGHT ENOUGH** to hold the saddle in place while mounting but loose enough to fit your fingers snugly underneath near the saddle flap.

3 **WHEN YOU PUT ON THE SADDLE**, tighten the girth gently, tighten it again when you leave the tie rack, and check it one more time prior to mounting.

4 **IF ONE END OF THE GIRTH** has elastic attachments near the buckle, that end is always buckled on the left side.

5 **DIP YOUR BIT IN A BUCKET OF WATER** after each ride and use a cloth or sponge to wipe away any slobber.

6 **WIPE YOUR EQUIPMENT DAILY** with a soft cloth to remove dust and mud.

7 **COVER EQUIPMENT** that is not being used so that it will not collect dust.

8 **IF YOUR HORSE IS TOO TALL TO BRIDLE,** hold the bridle by the cheek pieces when you introduce the bit to the horse's mouth.

English Tack

ENGLISH TACK WAS DESIGNED FOR GALLOPING THROUGH FIELDS and jumping over fences. The saddle was designed to give the horse the freedom to jump and to give the rider comfort and balance while jumping. It also allows the rider's leg to come into close contact with the horse's body.

ENGLISH SADDLE ANATOMY

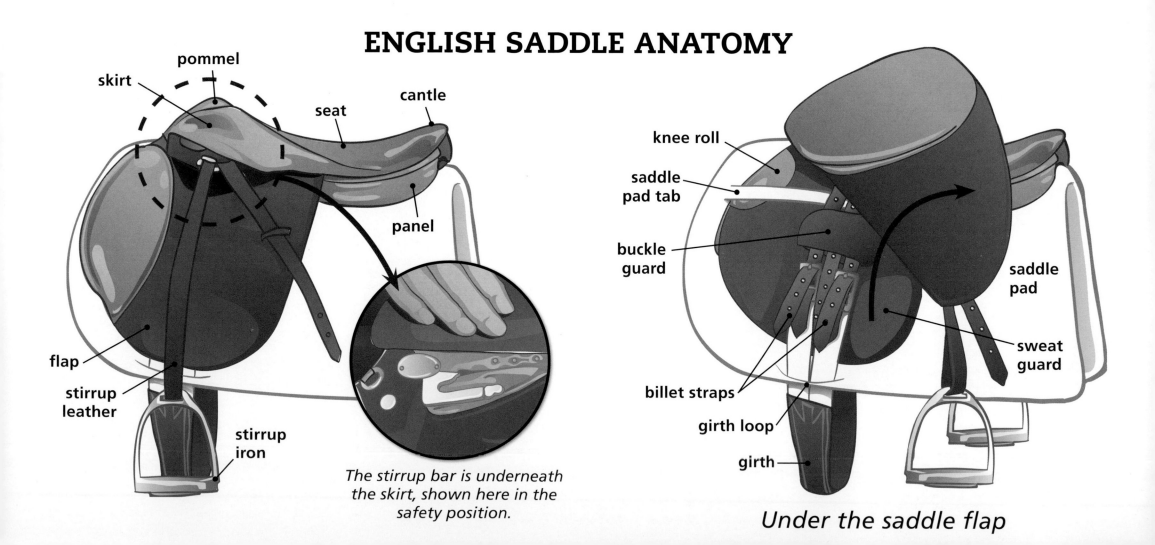

pommel
skirt
cantle
seat
panel
flap
stirrup leather
stirrup iron

The stirrup bar is underneath the skirt, shown here in the safety position.

knee roll
saddle pad tab
buckle guard
billet straps
girth loop
girth
saddle pad
sweat guard

Under the saddle flap

ENGLISH BITS

The snaffle bit puts pressure on the corners of the horse's mouth, tongue, and bars of the jaw. A horse with a snaffle bit is usually ridden with one rein in each hand. The smooth eggbutt snaffle is the basic bit for schooling a horse with an average mouth. The smaller the diameter of the mouthpiece and the rougher the texture, the more severe the pressure.

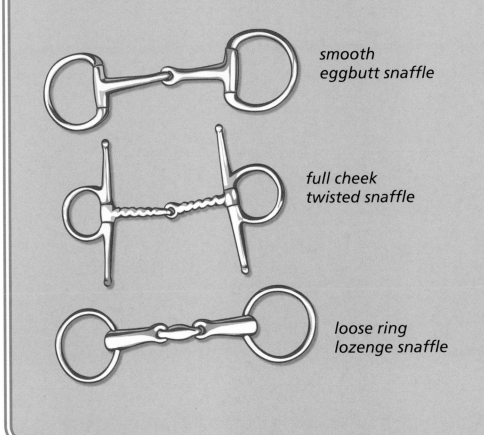

smooth eggbutt snaffle

full cheek twisted snaffle

loose ring lozenge snaffle

ENGLISH BRIDLE ANATOMY

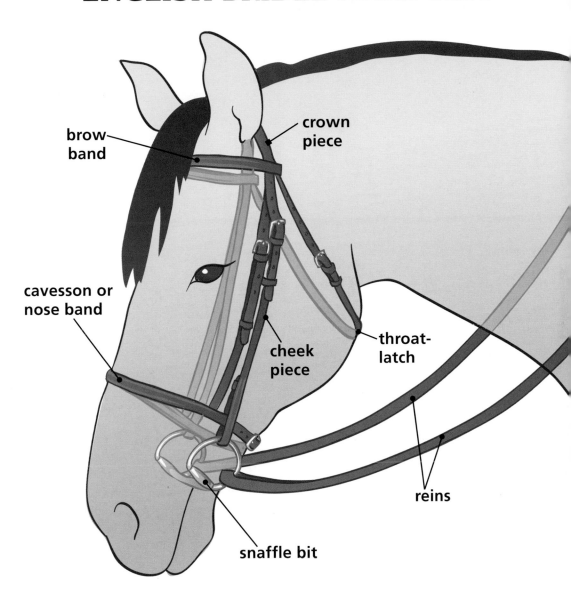

brow band

crown piece

cavesson or nose band

cheek piece

throat-latch

reins

snaffle bit

Saddling Up (English)

TO BEGIN, stand at the horse's left side, holding your saddle pad.

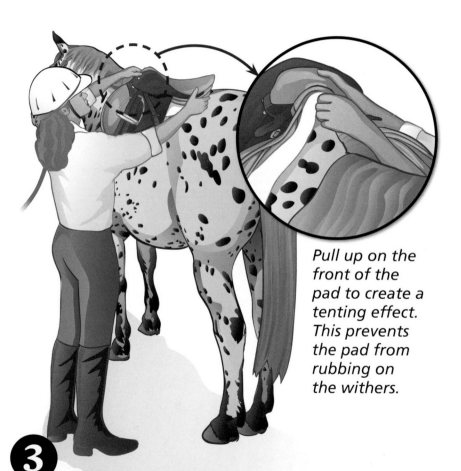

Pull up on the front of the pad to create a tenting effect. This prevents the pad from rubbing on the withers.

❶ If the pad is rectangular, fold it down the middle of the short side; if it is fitted, fold it evenly in half.

❷ Pinch the pad precisely at the fold and center it on the horse's back, with the front center of the pad resting over the withers.

❸ Hold the saddle with your left hand on the pommel and your right hand on the cantle. Place the saddle on top of the pad near the horse's withers with the pommel pointing forward.

4 With the girth in hand, walk around the front of the horse to the **right side** and fasten the saddle pad tabs under the billet straps.

5 Run the girth through the girth loop and fasten the buckles to the first and third billet straps.

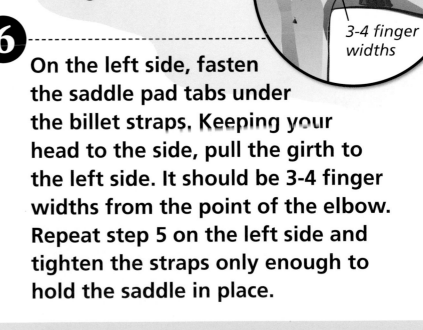

3-4 finger widths

6 On the left side, fasten the saddle pad tabs under the billet straps. Keeping your head to the side, pull the girth to the left side. It should be 3-4 finger widths from the point of the elbow. Repeat step 5 on the left side and tighten the straps only enough to hold the saddle in place.

PROPER SADDLE POSITIONS

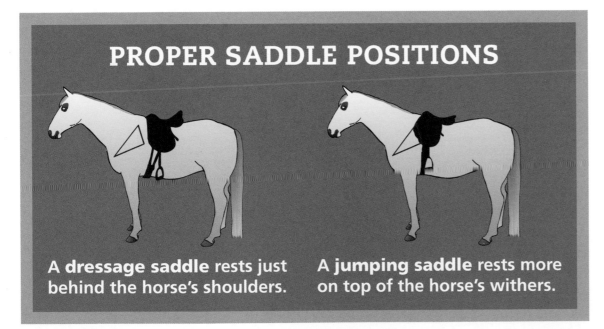

A **dressage saddle** rests just behind the horse's shoulders.

A **jumping saddle** rests more on top of the horse's withers.

Unsaddling (English)

THE LIGHT WEIGHT OF THE ENGLISH SADDLE makes it as easy to take off and carry as it is to put on the horse's back.

1 Stand facing the left side of the horse. "Run up" the stirrup by holding the front of the loop of leather and pushing up the stirrup iron until it touches the metal bar of the saddle.

2 Pull the leather through the stirrup iron so that the iron will not slide down.

3 Unbuckle the girth on the left side of the horse and allow it to gently swing behind the horse's front legs.

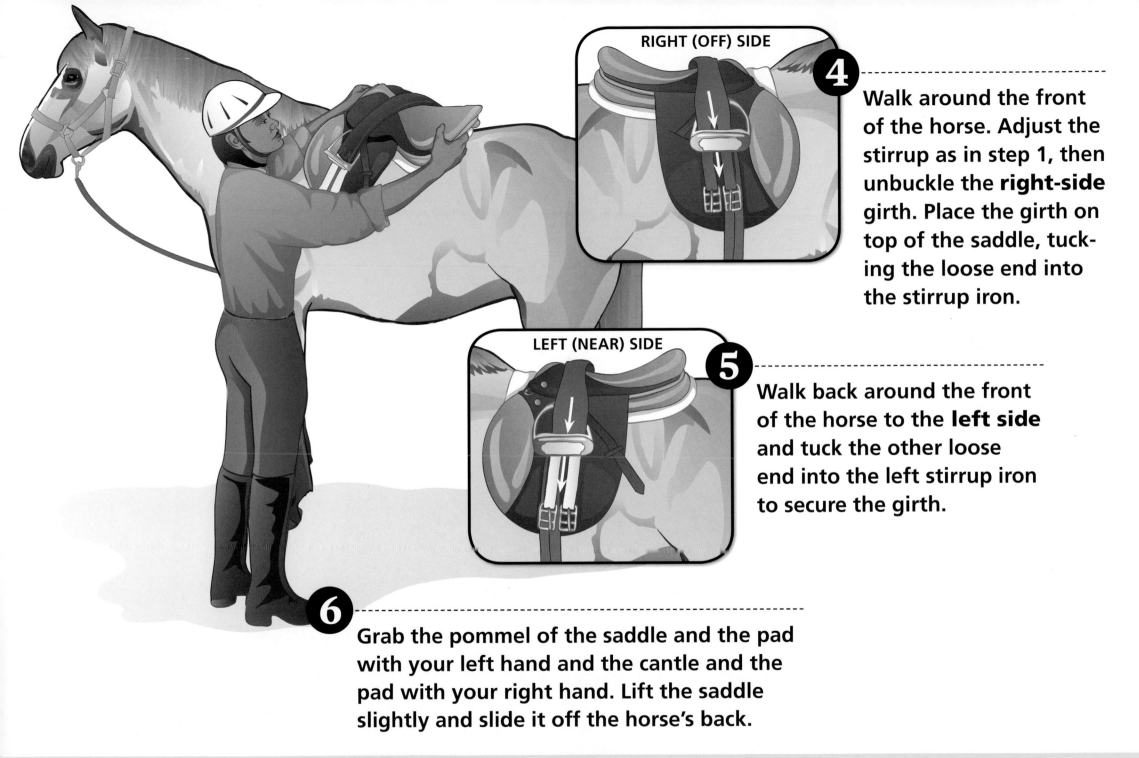

RIGHT (OFF) SIDE

4 Walk around the front of the horse. Adjust the stirrup as in step 1, then unbuckle the **right-side** girth. Place the girth on top of the saddle, tucking the loose end into the stirrup iron.

LEFT (NEAR) SIDE

5 Walk back around the front of the horse to the **left side** and tuck the other loose end into the left stirrup iron to secure the girth.

6 Grab the pommel of the saddle and the pad with your left hand and the cantle and the pad with your right hand. Lift the saddle slightly and slide it off the horse's back.

Bridling (English)

FOLLOW THESE INSTRUCTIONS for English bridling. When the halter is around the horse's neck, don't allow the horse's head to drop so low that he might step into the halter.

1 Standing on the horse's left side near his head, unsnap the lead rope or cross ties. Unfasten the horse's halter and then refasten it around his neck, so that he won't become loose.

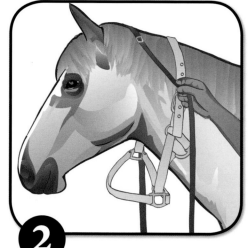

2 Put the buckle of the reins over his head so that the reins hang over the neck.

TIP

Put your right hand on the horse's poll and gently ask the horse to lower his head before you start to bridle him.

3 Hold the crown piece of the bridle in your right hand and place the thumb and first finger of your left hand along the width (see inset). Lift the bridle so that your right wrist rests between the horse's eye and his ear and the bit is near his mouth.

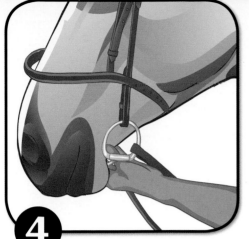

4 Lower the bridle and use your left hand to slip the bit under the horse's chin. Pull snug against the horse's chin.

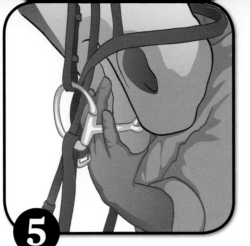

5 Slide the bit up to the horse's mouth, using the middle finger of your left hand to open his mouth on the **right side.** As the horse opens his mouth, slide in the bit and gently pull up with your right hand.

6 Hook the crown piece over the left and right ears, carefully bending each ear forward.

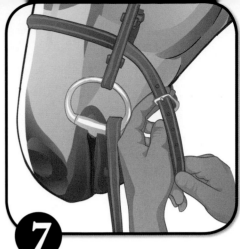

7 Fasten the cavesson snugly around the horse's muzzle, leaving enough room to slip one to two fingers between the cavesson and the jaw.

8 Fasten the throatlatch so that it hangs in the middle of the horse's cheek and is about three finger-widths from the throat.

TIP

Do not fasten the throatlatch tightly as it will put pressure on the horse's throat when he flexes his head at the poll.

Unbridling (English)

SOME HORSES ARE IN A HURRY to have their bridle taken off. It is important for your horse to be patient and respectful when waiting to be unbridled. Stand on the horse's left side, near his head, to unbridle him.

1 Fasten the horse's halter around his neck so that he is not in danger of becoming loose.

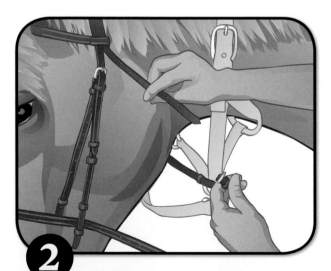

2 Unfasten all of the straps under the horse's jaw (namely the cavesson and the throatlatch).

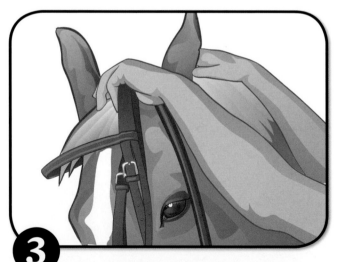

3 Pull the reins up to the crown piece of the bridle. Lift the bridle and reins over his ears, allowing him to slowly spit out the bit.

4 Put the bridle over your wrist and unfasten the halter, keeping the buckle in your left hand and the strap in your right.

5

Lower the halter to the horse's nose. Slip in the nose and raise the halter so that you can buckle it at the left ear.

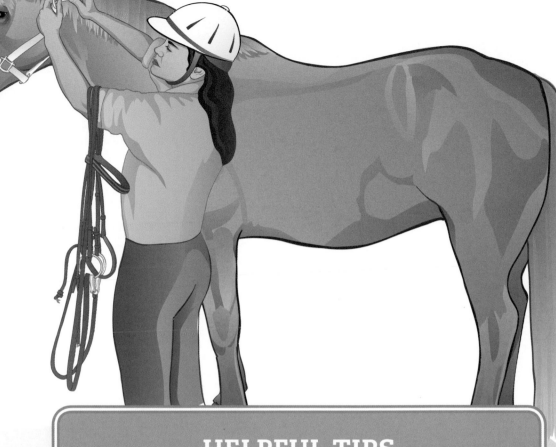

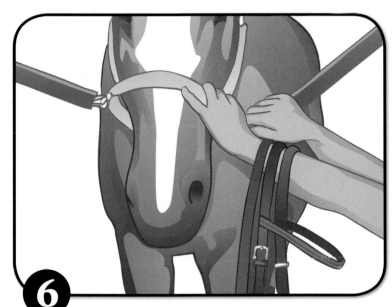

6

Snap the lead rope to the halter or cross-tie the horse.

HELPFUL TIPS

- A snaffle bit puts pressure on the corners of a horse's mouth; a curb bit, along with a curb strap, puts pressure under the horse's jaw.

- Do not pull the bit out of the horse's mouth when you unbridle him. Let the horse carefully release it.

Putting on a Bareback Pad

FOR PLEASURE RIDING AT A WALK, a bareback pad is a relaxing alternative to riding with a saddle. Use care when riding with a bareback pad as it tends to slip while mounting and riding.

1

Fold the pad down the middle so that the two girth straps touch. Pinch the pad precisely at the fold in the front and the back. Stand at the horse's left side and place the middle of the pad in the middle of the horse's shoulder, resting your fingers on the horse's mane and spine.

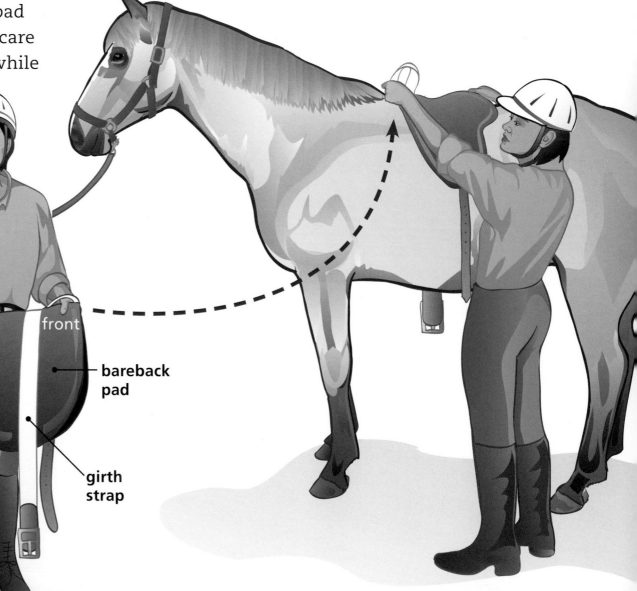

back

front

bareback pad

girth strap

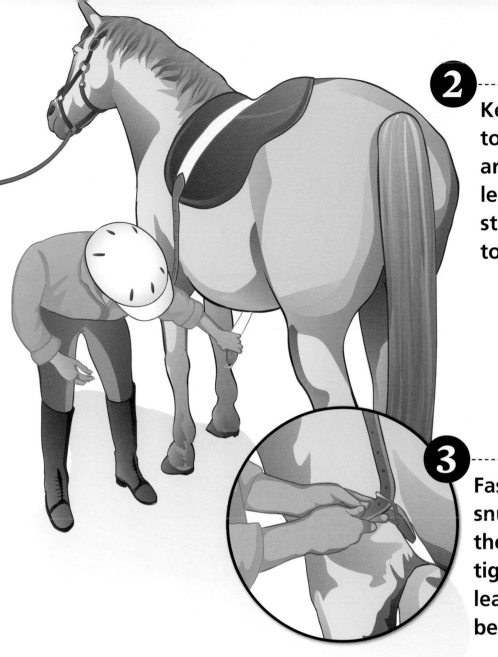

2

Keeping your head to the side, reach your arm behind the front leg and pull the girth strap with the buckle to the left side.

3

Fasten the girth strap snugly enough to hold the pad in place, then tighten it again when leaving the tie rail and before mounting.

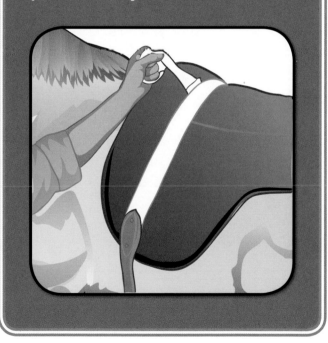

REMOVING THE BAREBACK PAD

Standing on the left side of the horse, unfasten the girth strap and let it swing gently behind the horse's front legs. Use your left hand to grab the front of the bareback pad and slide the pad toward you.

Western Tack

WESTERN SADDLES WERE DESIGNED FOR RIDING THE RANGE all day and roping cattle. The saddle distributes the rider's weight over a larger area of the horse's back. The Western bridle with the curb bit makes it easier to direct the horse with one hand.

WESTERN SADDLE ANATOMY

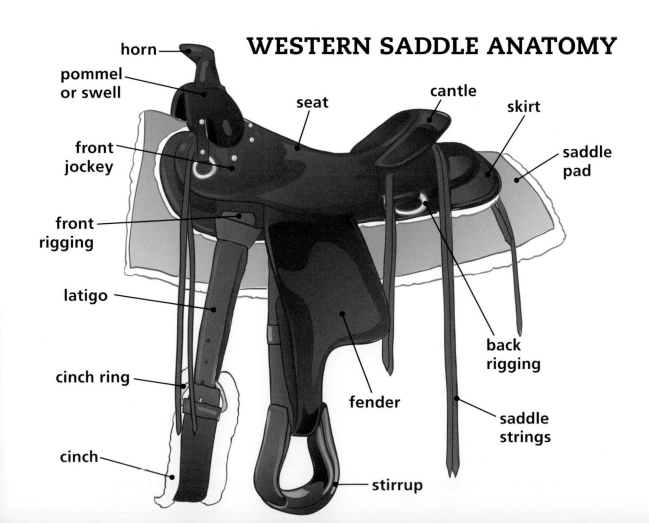

- horn
- pommel or swell
- front jockey
- front rigging
- latigo
- cinch ring
- cinch
- seat
- cantle
- skirt
- saddle pad
- back rigging
- fender
- saddle strings
- stirrup

TYPES OF CINCHES

When the Western cinch is tightened to hold the saddle in place, it must be wide enough to distribute the pressure evenly. The roper cinch is wider in the middle than the fleece-lined cinch to give more stability during rigorous activity, but the fleece-lined cinch is equally comfortable as long as it is kept clean.

Make sure to check the cinch before mounting because it is hard to tighten the cinch when you are mounted. If the saddle has a back cinch, it may be safer to unbuckle and remove it before pleasure riding.

fleece-lined western cinch

woven roper cinch

WESTERN BITS

The Western curb bit applies pressure to the tongue, bars of the jaw, chin groove, poll, and sometimes roof of mouth. The shanks act as a lever to create pressure on the chin and mouth via a curb strap. The longer the shank and the higher the port, the more severe the pressure.

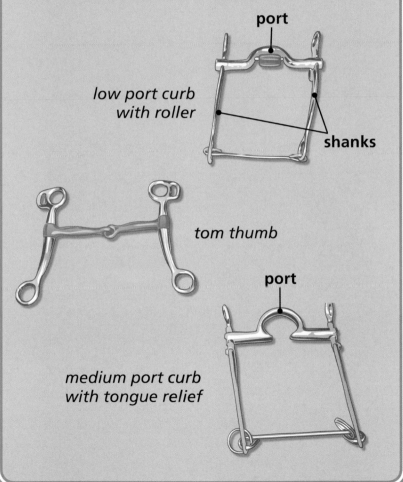

port

*low port curb
with roller*

shanks

tom thumb

port

*medium port curb
with tongue relief*

WESTERN BRIDLE ANATOMY

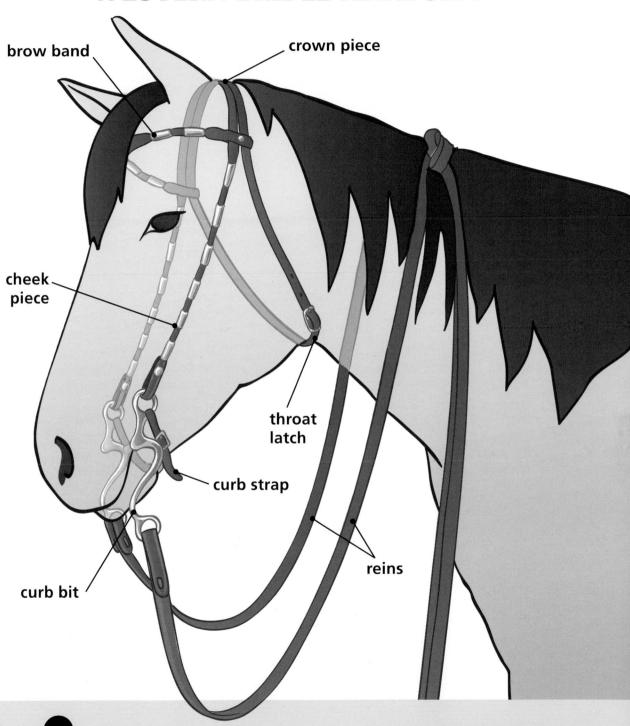

brow band

crown piece

cheek
piece

throat
latch

curb strap

curb bit

reins

Saddling Up (Western)

FOR THE COMFORT OF YOUR HORSE, make sure his saddle fits well. When the saddle is sitting on the horse's back without a pad, you must be able to fit at least two fingers between the front of the saddle and the horse's withers. When saddling, use a pad to help cushion the horse's back.

1

Standing at the horse's left side, fold your rectangular saddle down the middle of the short side. Pinch the pad precisely at the fold and place it on the horse's back, resting your fingers on his mane and spine. Place the front of the pad in the middle of the horse's shoulder.

PROPER SADDLE POSITION

A Western saddle rests just behind the shoulder to allow freedom of movement in the shoulder.

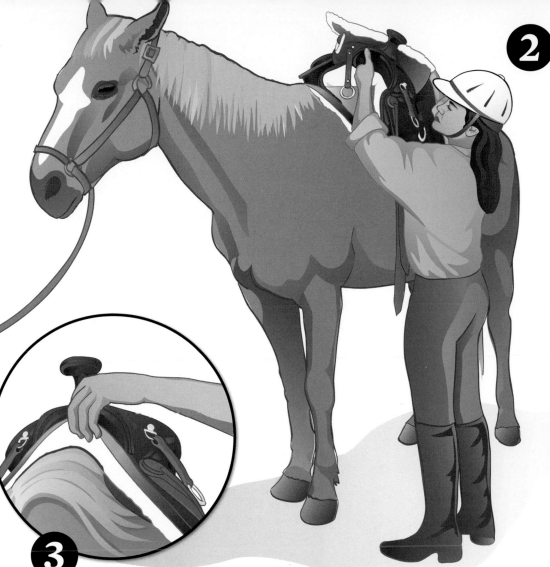

2 Hold the saddle with your left hand near the horn and your right hand on the cantle with the right stirrup and the cinch flipped onto the top of the saddle. From the horse's left side, gently center the saddle on the pad with the horn pointing forward.

3 Allow the saddle and the pad to slide back so that they rest comfortably behind the horse's shoulder blade. Pull up the pad under the pommel to create a tenting effect.

4 Walk around the front of the horse to the right side and pull down the stirrup and the cinch. Walk back around to the left side. Keeping your head to the side, reach under the horse behind the front leg and pull the cinch to the left side (as shown).

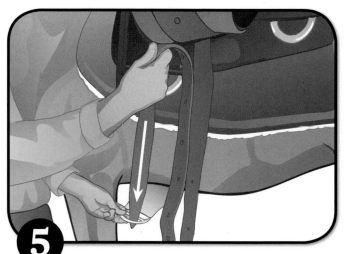

5 Place the left stirrup on the horn of the saddle. Run the latigo through the ring of the cinch from the top down.

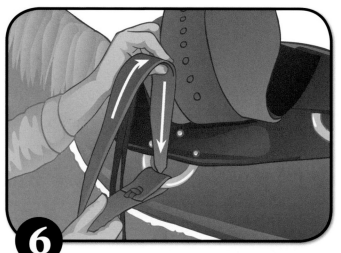

6 Gently pull up the latigo and run it through the rigging ring it was originally attached to. Repeat this process one more time through the cinch ring and the rigging ring.

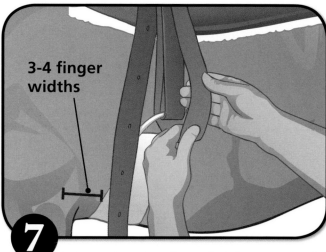

3-4 finger widths

7 The cinch should be roughly 3–4 finger widths back from the point of the elbow. Pull the slack out of the latigo so that the cinch rests snugly against the horse's body.

8 Pull the excess latigo out to the left side of the rigging ring.

TIP

You should count 4 thicknesses of latigo stacked on top of each other between the cinch and the rigging ring.

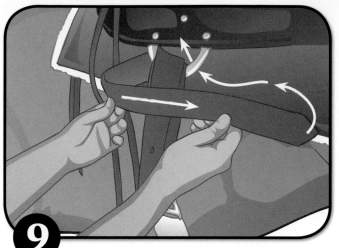

9 Fold the excess latigo to the right, laying it across the stacked latigo. Thread the tip through the rigging ring from underneath.

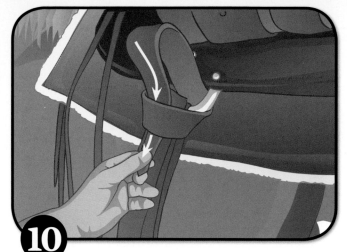

10 Run the loose end down through the horizontal piece of latigo to lock it in place, much as you would tie a men's tie.

11 To tighten the cinch, pull the outermost layer of stacked latigo toward you, then work the slack through the knot so that the stack and knot are flat. Place the excess latigo in the slot on the saddle, near the horn, or through the back rigging ring.

CHECKING THE CINCH

- Before mounting, tighten the cinch gradually, allowing your horse to walk a few steps each time you pull up the latigo. The cinch should be tight enough to hold the saddle in place while mounting.

- If the cinch has a buckle and the latigo has holes, you can buckle the latigo into the cinch. You don't need to tie a knot around the rigging ring.

Unsaddling (Western)

AFTER YOU TAKE YOUR SADDLE OFF, make sure that your pad is left in an area where the underside can dry before your next ride.

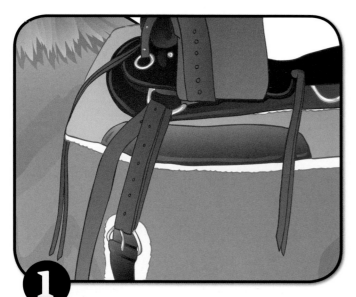

1 Stand facing the left side of the horse. Place the left stirrup on the horn and untie the latigo knot around the rigging ring. The loose end of the latigo should be hanging near the four layers of leather connected to the cinch.

2 Grab the top piece of leather that connects the cinch to the rigging with both hands and tug sharply to loosen the latigo. Unwrap the latigo between the cinch and the rigging and let the cinch gently swing behind the horse's front legs.

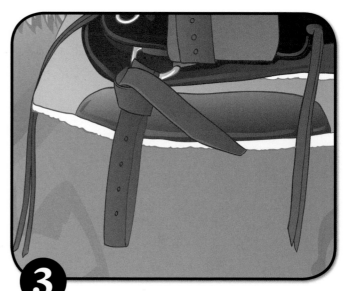

3 Run the loose latigo strap back through the rigging ring several times, then wrap the latigo around itself and secure it.

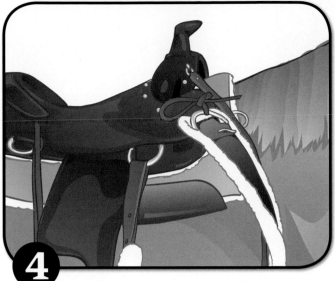

4 Walk around the front of the horse to the right side. Attach the cinch ring to the saddle using the saddle strings (shown above) or simply lay the cinch over the top of the saddle.

NOTE

To carry your saddle in one hand, slip your fingers in the hole just behind the horn and lift the saddle onto your hip, with the horn facing you. If you have to set your saddle on the ground, tip it up so that the front of the horn touches the ground and the saddle rests on the front edges of the skirt.

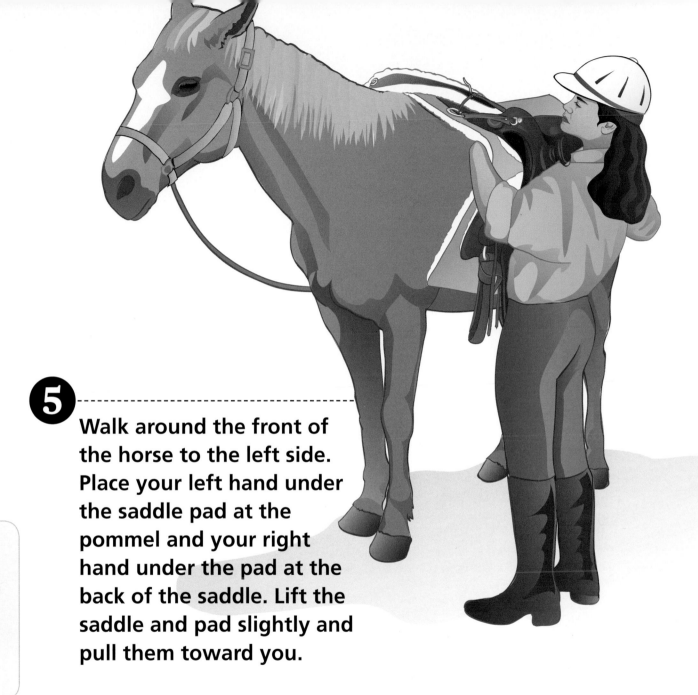

5 Walk around the front of the horse to the left side. Place your left hand under the saddle pad at the pommel and your right hand under the pad at the back of the saddle. Lift the saddle and pad slightly and pull them toward you.

Bridling (Western)

MAKE SURE THAT YOUR BRIDLE IS FITTED PROPERLY so that it is comfortable for your horse and goes on with ease. To begin, stand on the horse's left side near his head and unsnap the lead rope or cross-ties.

1 Unfasten the horse's halter then refasten it around the horse's neck, so that the horse is not in danger of becoming loose. Hang the reins over the horse's neck.

2 Hold the crown piece of the bridle in your right hand. Lift the bridle to the horse's head so that your right wrist rests between the horse's eye and his ear and the bit is near the horse's mouth.

3 Lower the bridle and use the left hand to slip the bit under the horse's chin. Pull it up with the right hand so that the bit is snug against the horse's chin.

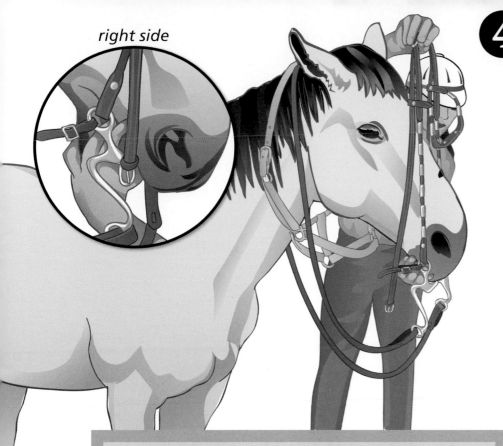

right side

4 Place the thumb and forefinger of your left hand along the width of the bit. Slide the bit up to the horse's mouth, using your middle finger to open the horse's mouth on the **right side** (see inset). As the horse opens his mouth, slide in the bit and pull up with the right hand, allowing the curb strap to slide under the chin.

HELPFUL TIPS

- Always untie your horse before bridling.

- Proper fit for a snaffle bridle allows that bit to rest in the corners of the mouth, producing two small wrinkles. Proper fit for a curb bridle allows the bit to rest in the corners of the mouth, producing one small wrinkle.

5 Hook the crown piece over the left and right ears, carefully bending each ear forward.

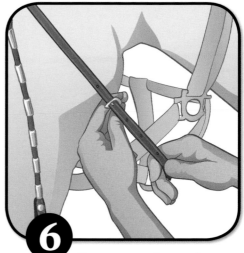

6 Fasten the throatlatch so that it hangs in the middle of the horse's cheek and is about three finger widths from the throat.

Unbridling (Western)

DO NOT LET YOUR REINS DROP TO THE GROUND while unbridling. If the reins do drop, collect them quickly so that the horse does not step on them. If he does so, he may pull back, break the reins, and injure his mouth.

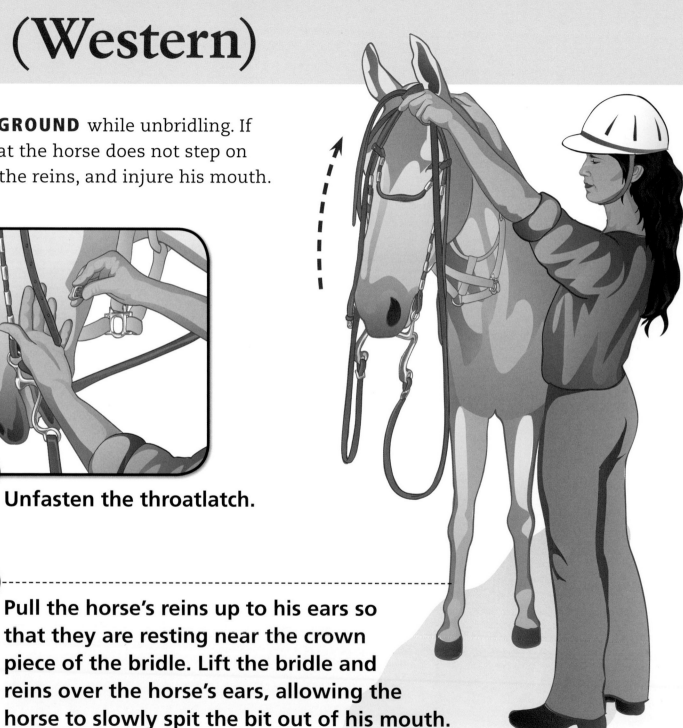

1 Standing on the horse's left side, fasten his halter around his neck so that he is not in danger of becoming loose.

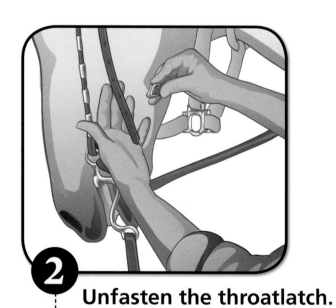

2 Unfasten the throatlatch.

3 Pull the horse's reins up to his ears so that they are resting near the crown piece of the bridle. Lift the bridle and reins over the horse's ears, allowing the horse to slowly spit the bit out of his mouth.

4 Put the bridle over your arm and unfasten the halter, keeping the buckle in your left hand and the strap in your right. Lower the halter to the horse's nose, slip in the nose and raise the halter so that you can buckle it at the left ear.

HELPFUL TIP
Wipe down your bit or dip it in water after each ride to keep it clean.

5 Snap the lead rope to the halter or cross-tie the horse.

MOUNTING TIPS

ALWAYS MOUNT FROM THE LEFT SIDE OF THE HORSE. When mounting large horses, you may want to use a mounting block (or a good-sized rock if you are out on the trail), or receive a lift from a friend, called a "leg up."

1 **ALWAYS WEAR A HELMET** when mounting and riding.

2 **ALWAYS CHECK YOUR GIRTH** before mounting to make sure it is tight enough.

3 **IF YOU TIGHTEN THE GIRTH BEFORE MOUNTING,** make sure to walk the horse a couple of steps before you put your foot in the stirrup and mount.

4 **TRY NOT TO STICK YOUR TOE INTO THE SIDE OF THE HORSE** after you put your foot in the stirrup. Point your toe forward as you step up on the horse.

5 **WHEN YOU STEP UP ON THE HORSE,** step toward the horse's head instead of stepping toward the other side.

6 **IF YOU FALL OR MUST DISMOUNT IN AN EMERGENCY,** make sure that you are clear of the reins and all equipment.

7 **WEAR BOOTS OR SHOES WITH A HEEL** when riding with stirrups.

8 **NEVER TIE YOUR HORSE TO ITS BIT OR BRIDLE.** Use a halter.

Mounting

STAND ON THE LEFT SIDE OF THE HORSE, facing his side. Always check your girth before mounting to make sure it is tight enough.

TIP

Some Western riders prefer to face the front of the horse, grab the reins in the left hand and the horn in the right hand, then put their left foot in the stirrup and swing on.

①

FOR ENGLISH, with your right hand grab the buckle on the reins above the withers and with your left hand grasp the two reins in front of the buckle.

FOR WESTERN, with your right hand grab the ends of the reins above the withers and with your left hand grasp the two reins in front of your right hand.

THEN, slide your left hand up the reins so that they are short enough to stop the horse if he moves.

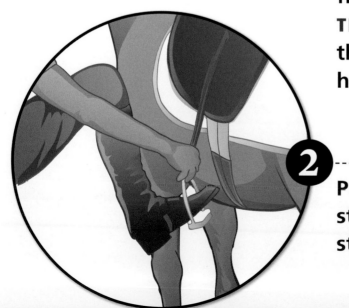

②

Put your left foot in the left stirrup while steadying the stirrup with your right hand.

TIP

If your horse is tall and there is no mounting block, let the left stirrup down several holes and mount from the ground. Shorten the stirrup after you have mounted.

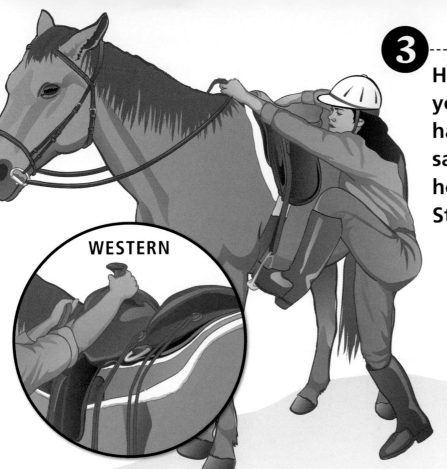

3

Hold the mane and reins with your left hand while your right hand reaches for the seat of the saddle (for English) or for the horn of the saddle (for Western). Step up on to the horse.

WESTERN

TIP

After your foot is in the stirrup, move around so that your right hip is close to the horse before stepping up.

TIP

As you step up on the horse, look toward the horse's ears so that you step toward the horse's head.

4

Swing your right leg over the top of the horse and land very gently in the saddle. Place your right foot in your right stirrup and take the reins in two hands.

Proper Position in the Saddle

PROPER POSITION IN THE SADDLE IS CRUCIAL TO GOOD RIDING. All of these styles align your ear, shoulder, and hip, but your foot placement and the length of the stirrup varies.

HUNTSEAT SADDLE

Proper position for a huntseat saddle means the straight line through your ear, shoulder, and hip meets the *back* of the heel of your boot.

DRESSAGE SADDLE

proper position for a dressage saddle means the straight line through your ear, shoulder, and hip meets the *middle* of the heel of your boot.

WESTERN SADDLE

Proper position for a western saddle is similar to the position for a huntseat saddle — the straight line from your ear to your hip meets the *back* of the heel of your boot.

GETTING A LEG UP

To get a "leg up," follow steps 1 through 3 of Mounting (pages 94–95), then have your helper grasp your bent left knee. On the count of three, your helper will lift your left knee toward the flap of the saddle while you jump with your right leg and swing it over the top of the horse.

Dismounting

ALWAYS DISMOUNT FROM THE HORSE'S LEFT SIDE.
Step 1, where your hands are placed at the outset, will be different for English and Western saddles. Steps 2 through 4 are the same for English and Western.

ENGLISH

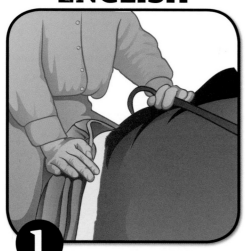

WESTERN

① FOR ENGLISH, place both reins in your left hand and rest that hand on the pommel of the saddle or on the withers of the horse. Place your right hand on the saddle to the right of the pommel. FOR WESTERN, take your reins in your left hand and rest that hand on the horn of the saddle. Place your right hand on the pommel of the saddle to the right of the horn.

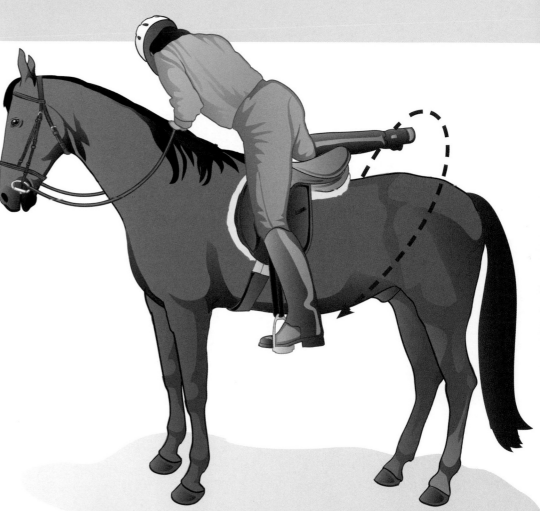

② Stand up in your stirrups over the horse. Take your right foot out of the stirrup and swing your right leg over the saddle and the horse's hindquarters.

3 Lean your right hip against the saddle so that all of your weight is resting on the horse and the saddle, then take your left foot out of the stirrup.

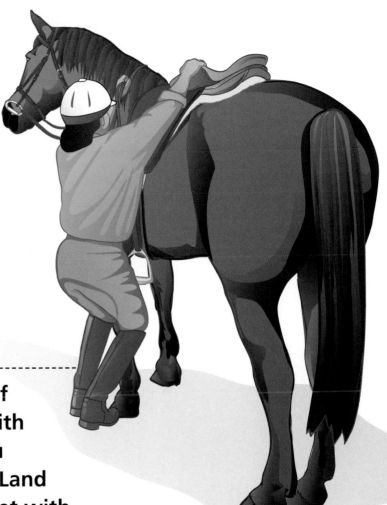

HELPFUL TIPS

- Try not to let your right leg touch the horse as you swing it over the back of the saddle.

- If the horse will not stand still, take both feet out of the stirrups at the same time and swing your right leg over the saddle as you lower yourself down.

4 Slide down that side of the horse, releasing with your right hand as you approach the ground. Land on the balls of your feet with your knees bent.

Emergency Dismount

AN EMERGENCY DISMOUNT IS USUALLY A LAST RESORT done to minimize injuries, but it can be dangerous. Do an emergency dismount when you think that you are going to fall anyway, or there is a loose horse coming your way and you don't have time to dismount and hold your horse.

1 Kick both feet out of the stirrups and make sure they are clear. Lean forward so that your chest is close to the horse's neck and release the reins.

2 Wrap your arms around the horse's neck as you move your legs parallel to the horse's back.

3 Choose the side of the horse that you wish to dismount (in an emergency it is going to be the side where you have already lost your balance) and use the arm that is on top of the horse's neck to slow the descent of your legs down the side of the horse.

4 If the horse is moving, bend your knees and start running before your feet hit the ground. As your feet touch the ground, let go and push away from the horse and run alongside the horse.

CAUTION

Do not try to hold the horse if he is moving. Make sure the reins are clear of your hands.

5 If you lose your footing and begin to fall, do not try to brace yourself with your hands. Roll into a ball, tucking in your head, and roll away from the horse.

HORSE BEHAVIOR

HORSES ARE VERY EXPRESSIVE ANIMALS that live in a herd and seek companionship. They want to bond with us and perform for us. The more you get to know your horse, the easier it is to read his mood and assess his well-being.

1 **THE BEST WAY TO GET TO KNOW YOUR HORSE** is to spend time grooming him and observing him in the stall.

2 **A HORSE'S HEARING** is his most acute sense. This is why horses find loud noises troubling and will respond differently to different tones of voice.

3 **A TIED HORSE THAT IS PAWING** the ground is impatient, but a stalled horse that is pawing the ground could be ill.

4 **HORSES ARE "FLIGHT, NOT FIGHT" ANIMALS** that choose to flee from anything that they perceive is dangerous. The more confidence your horse has in you, the more likely he is to override his instinct.

5 **HORSES ARE ACCUSTOMED TO TAKING INSTRUCTION FROM THE BOSS MARE** of the herd, so when you assert yourself and gain your horse's respect, the horse feels secure and included in your herd.

6 **HORSES WILL CURL THEIR UPPER LIP** in response to an unusual smell, taste, or intense pain. This behavior is called the Flehmen response.

7 **A HORSE'S TAIL CAN SIGNAL HIS MOOD.** A horse will swish his tail dramatically to express irritation and raise it over his back like a flag when he is excited.

Body Language

HORSES PRIMARILY USE NONVERBAL FORMS OF COMMUNICATION.
Their eyes, ears, body tension, and posture indicate what they are feeling.
As with all prey animals, if one horse reacts dramatically to a stimulus,
like a loud noise, other horses will react similarly. This instinct helps them
survive in the wild.

EARS PRICKED FORWARD can be a sign of curiosity or, combined with a wide eye, fear.

EARS RELAXED forward, to the sides, or leaned back, combined with a soft eye, are a sign of contentment.

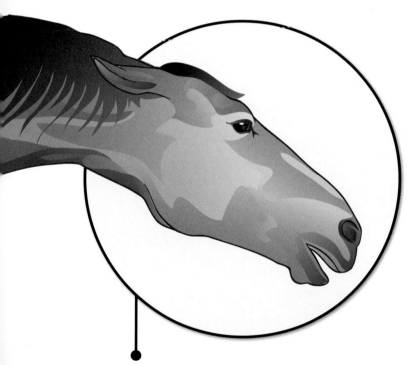

EARS LAID BACK FLAT, combined with slanted and steeled eyes, are a sign of aggression.

Vocalizing

ALTHOUGH THEIR COMMUNICATION IS PRIMARILY PHYSICAL, horses do occasionally vocalize audibly to one another.

IN THE SAME WAY that a mare will nicker to her foal, your horse might nicker to you.

WHEN HORSES MEET for the first time, they breathe heavily into each other's nostrils, bow up their necks, and sometimes squeal and strike the ground with a front foot.

HELPFUL TIPS

- Be careful when you let two horses touch noses, because they might squeal and strike with a front foot and possibly hit you.

- Horses will vocalize when they think another horse is in trouble. For example, when a horse is scrambling in a trailer or is "cast" in the stall, many others will whinny loudly to express their concern.

- Some horses will whinny just to communicate with others in the area, especially when entering a new environment.

- A stallion makes a distinctive low guttural grunting when he is becoming sexually excited. If you are handling a horse who begins to make these sounds, walk him away from other horses until he calms down.

Reading Your Horse's Mood

YOU CAN LEARN TO INTERPRET the signs that horses give each other to show what they are feeling. Some of these signs are subtle, like the flare of a nostril when a horse is fearful, while others, like the long snakelike neck and head of an unfriendly horse, are extremely obvious.

MOOD		EARS	EYES	NOSTRILS	BODY TENSION
Content		Relaxed and leaned back, to the sides, or gently forward	Soft and round, possibly sleepy looking	Elongated and relaxed	Almost none; relaxed
Friendly		Relaxed but tilted forward	Round and alert with a happy expression; focused on what is approaching	Elongated and relaxed	Some slight body tension, but in a flexible, welcoming posture

Unfriendly	Laid flat back, pressed against the neck	Squinting and slanted with a hard, cold look	Pulled tightly back and wrinkled; almost snarling	Stiff, especially the neck
Nervous	Tilted forward or to the sides	Slightly almond-shaped with a tense expression	Round and tense	Tense and ready to move at any moment
Fearful	Pricked forward	Wide, round, and alert with a tense expression	Flared wide	Tense and rigid, ready to flee at any moment

HORSES EXHIBIT A VARIETY OF SIGNS THAT THEY ARE AGITATED and could react violently. Their bodies become tense, their eyes look strained, and their general demeanor appears upset. Their instinct is to flee or bolt, but if they are restrained and cannot leave the situation, they may act out violently.

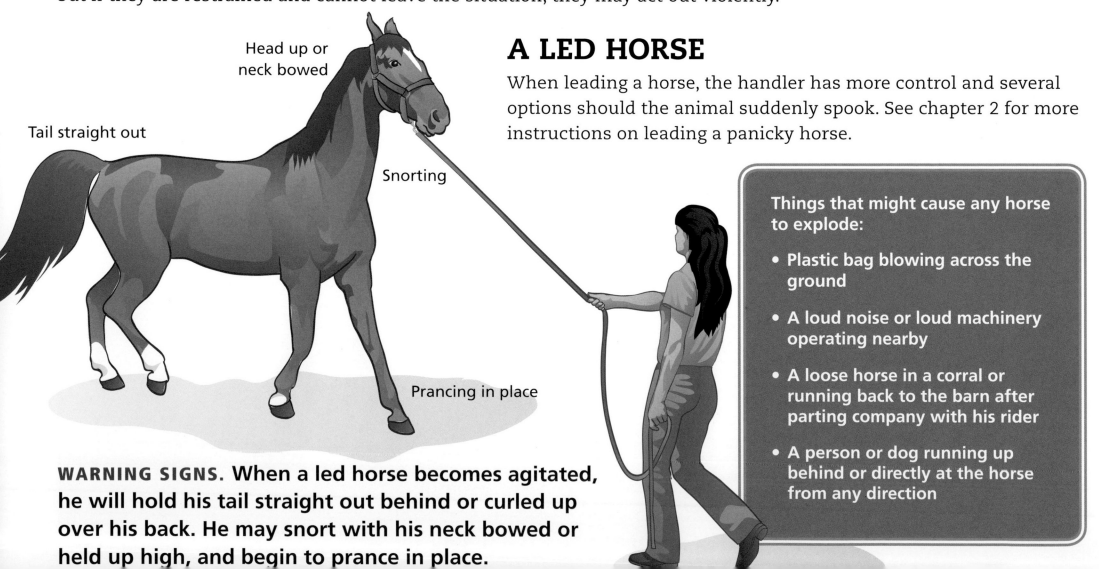

Head up or neck bowed

Tail straight out

Snorting

Prancing in place

A LED HORSE

When leading a horse, the handler has more control and several options should the animal suddenly spook. See chapter 2 for more instructions on leading a panicky horse.

Things that might cause any horse to explode:

- Plastic bag blowing across the ground

- A loud noise or loud machinery operating nearby

- A loose horse in a corral or running back to the barn after parting company with his rider

- A person or dog running up behind or directly at the horse from any direction

WARNING SIGNS. When a led horse becomes agitated, he will hold his tail straight out behind or curled up over his back. He may snort with his neck bowed or held up high, and begin to prance in place.

BEFORE A LED HORSE EXPLODES

- Walk the horse in a circle.

- Use the Emergency Lip Rope (page 19) to lead him safely.

- Apply a skin twitch.

- Lead him back to the stall and turn him loose so that he can calm down.

WHEN A LED HORSE EXPLODES

- **IF HE REARS,** do not pull on the lead while the horse is in the air. Instead, raise your hand with the lead rope in it and keep yourself clear of the front feet. Make sure that there is slack in the rope until the horse is back on the ground, then walk him in a circle.

- **IF THE HORSE RUNS BACKWARD,** do not pull on the lead rope. Move with the horse and keep a little bit of slack in the rope. When he stops, reassure him with a pat on the neck and lead him forward.

- Lead the horse back to the stall and turn him loose so that he can calm down.

NOTE

To apply a skin twitch, grab a handful of skin where the horse's shoulder meets the neck. Give your hand a quarter turn to get the horse's attention and minimize his fractious behavior.

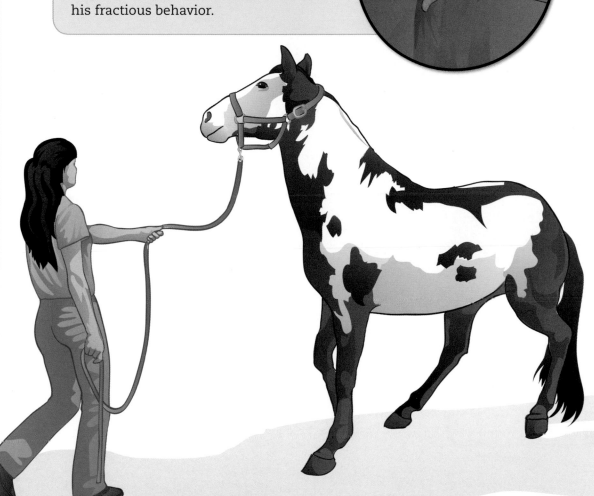

IF THE HORSE RUNS BACKWARD, move with him and keep a bit of slack in the rope.

A TIED HORSE

Signs of an imminent explosion include pulling against the rope, an inability to stand still, hopping from foot to foot, and general body tension with a bowed or stiff neck and wide eyes.

BEFORE A TIED HORSE EXPLODES

- If the horse seems about to spook, untie him and remove him from the stimulus.

- If he does not calm down, lead him directly to his stall to relax.

- Use the Emergency Lip Rope (see page 19) if the horse is really out of control.

WHEN A TIED HORSE EXPLODES

- Move away from him so you will not be in the way of flailing feet.

- Wait until he calms down before approaching. Even then approach carefully.

- Carefully unsnap the cross-ties if he is cross-tied. If he is single-tied, unsnap the lead rope from the halter and attach a separate lead rope, which you hold.

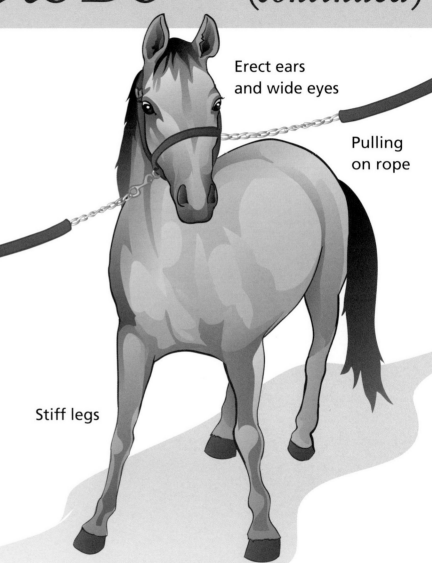

Erect ears and wide eyes

Pulling on rope

Stiff legs

WARNING SIGNS. A rigid posture and wide eyes may be a warning that the horse is about to explode.

Human Body Language

HORSES ARE EXPERT AT READING BODY POSTURE, not only in other horses, but also in humans. Tone of voice is also important, as are the use of your eyes and the tension of your body.

TO A HORSE, who reads body language, seeing you approach with your shoulders relaxed rather than squared can be as meaningful as hearing you speak in a normal tone rather than screaming. A predator would approach his prey with his shoulders squared, and since a horse is a prey animal, this posture is likely to make him uneasy.

MIRRORING YOUR MOOD

Horses will take their cue about how to react from their handlers. If you are calm and in control, it is more likely that you will be able to keep your horse calm. If you are tense, the horse will sense that tension and mirror it.

Approaching a horse with your shoulders relaxed and eyes cast downward is a nonthreatening posture.

Approaching a horse with squared shoulders, staring right in his eye, signifies dominance and possibly aggression.

YOUR HORSE, YOUR FRIEND

HORSES ARE SENSITIVE ANIMALS THAT HAVE EMOTIONS AND FEELINGS. If you try to meet their emotional needs, as well as their physical needs, you enhance their desire to please you. Wild horses play together as youngsters and adolescents, and this sense of fun can be incorporated into your relationship with them.

1 **IF YOU ACCIDENTALLY SCARE OR ANNOY YOUR HORSE,** apologize with a reassuring pat and kind words so that he will understand that it was unintentional.

2 **FOOD IS AN EASY WAY TO WIN YOUR HORSE'S AFFECTION,** but he needs to respect you, because with respect comes true companionship.

3 **ALTHOUGH IT IS TEMPTING, DO NOT LET YOUR HORSE SEARCH YOUR POCKETS FOR TREATS.** This can lead to a dangerous habit of biting your clothes and knocking you over.

4 **IF YOUR HORSE IS TURNED OUT,** following you is a sign of affection and respect, but playing tag is dangerous because he may treat you like another horse.

5 **TEACHING YOUR HORSE TRICKS** is a way to keep him entertained and interested when he can't be ridden due to injury or weather.

6 **IF YOUR HORSE DOES NOT GET ENOUGH ATTENTION FROM YOU** or time out of his stall, he can develop vices like cribbing and weaving in the stall.

7 **HORSES HAVE GOOD MEMORIES,** and past experiences affect how they react in the present. Positive situations and experiences produce a confident and responsive horse.

8 **RELATIONSHIPS ARE SO IMPORTANT** to horses that when two live together they can become so attached that neither can relax when they are apart.

Rewarding Your Horse

HORSES PREFER A FIRM, FRIENDLY TOUCH to one that is overly light and delicate. Horses touch each other with their noses and lips. Make your hand feel like the pressure of a horse's nose when you touch and greet him.

OFFER A TREAT in the center of your palm with the hand held open and fingers flat.

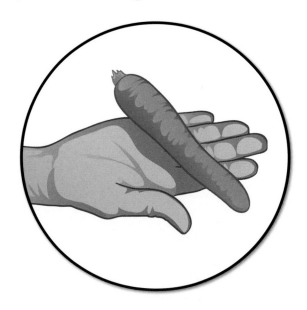

YOU CAN MAKE A HORSE A FRIEND by rubbing the base of his withers, near where the mane ends, with your knuckles. When you hit the right spot, the horse will stick out his nose and try to "rub" the air in appreciation.

CAUTION

Some horses will rub you with their nose in response to having their withers rubbed. If this happens, use caution, as they often bite as well.

Basic Tricks

HORSES CAN LEARN A VARIETY OF TRICKS, especially when there is a tasty reward at hand. To teach a horse a trick, you must break the trick down into very small steps.

GIVING A SMILE

Attach a halter and a lead rope to your horse but don't tie him.

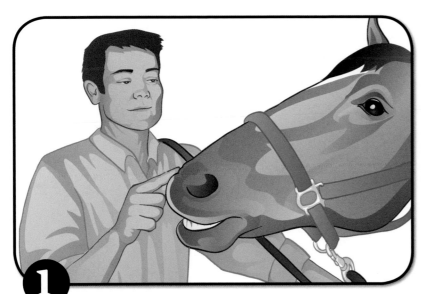

1 Lightly touch your horse's nose with your index finger. Point toward the sky and give the command "smile." When the horse moves his upper lip, even slightly, reward him with a treat.

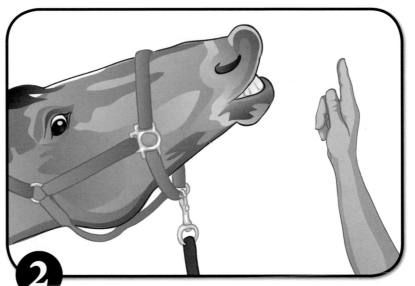

2 Repeat this process, rewarding any lip movement at first, then waiting to reward greater progress. Eventually, when you point to the horse then point in the air, the horse will smile.

Basic Tricks (continued)

GIVING A HUG

Attach a halter and lead rope but don't tie your horse. First, you must train the horse to move his head to any place where you put your hand. To accomplish this, do the following:

1 Hold a small treat like a bit of carrot or grain in your closed fist and hold it up to the horse's nose. Touch the horse's nose, give the command "here," and then feed him the treat. Do this several times.

2 Move your fist that contains a treat a short distance away from the horse's head and give the command "here." When the horse touches your closed fist, hand him the treat. Repeat this process until the horse is touching your closed fist in different locations around his head.

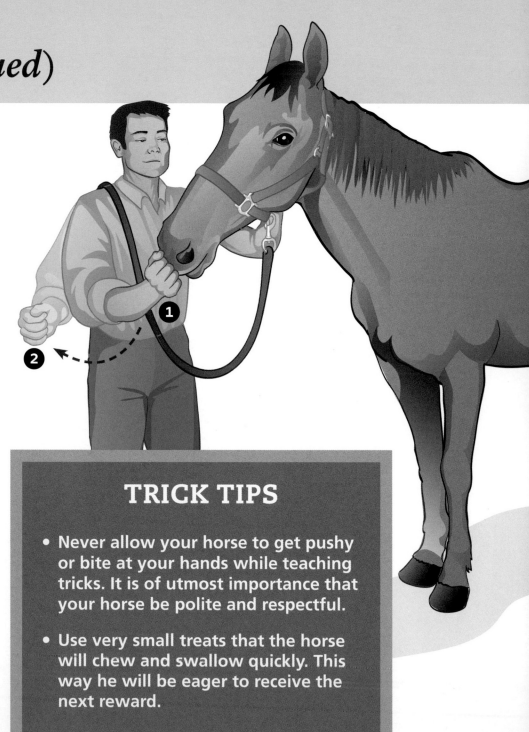

TRICK TIPS

- Never allow your horse to get pushy or bite at your hands while teaching tricks. It is of utmost importance that your horse be polite and respectful.

- Use very small treats that the horse will chew and swallow quickly. This way he will be eager to receive the next reward.

3

Stand at the horse's left side, near his shoulder, facing the rear. Place your left arm around his chest and have a treat in your closed right fist. Say "give me a hug" as you reach behind you toward his head. Repeat this process several times, each time moving your right fist closer to your side.

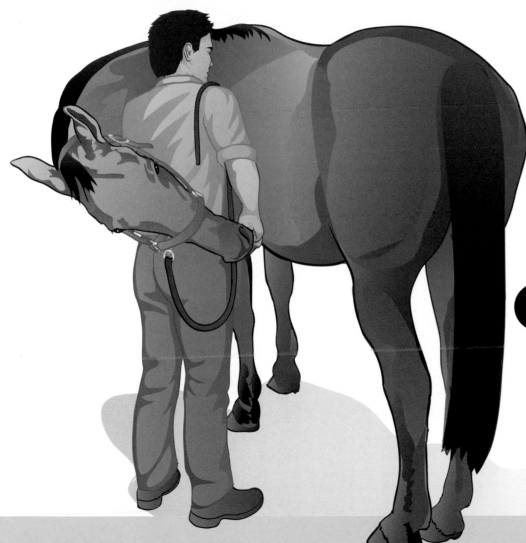

4

Soon, when you put your left arm around the horse's chest and say "give me a hug," he will wrap his neck around your body.

GIVING A BOW

Attach a halter and a lead rope to the horse but don't tie him. Teach this trick after you have taught your horse how to give you a hug.

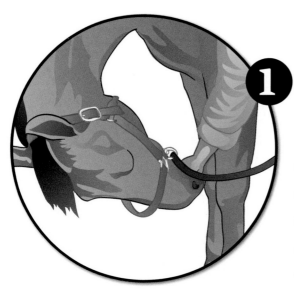

1 Reach down toward his knees with a treat in your closed fist and get him to consistently touch your fist down by his knees. (He'll do this if he's already been taught to hug.)

2 Stand at the left side and reach between the horse's front legs toward his head with a treat in your closed fist. Give him the command "here," so that his nose touches your fist. Repeat this process, each time moving your hand farther back between his front legs.

CAUTION
Be careful that the lead rope does not get under your feet or your horse's hooves while you are practicing tricks.

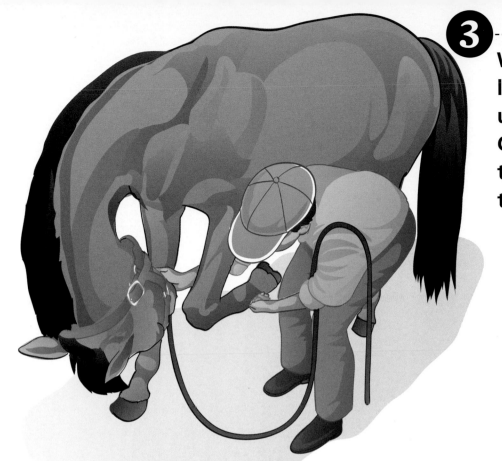

With your closed right fist between his front legs, tap his left fetlock with your foot and then use your left hand to pick up the left front leg. Give the command "bow," then give him the treat when he touches your closed fist. Repeat this process slowly several times.

4

Soon when you tap his left fetlock with your foot and bend down with your right closed fist, your horse will be bowing for a treat.

Grooming for a Show

ALLOW ENOUGH TIME TO PREPARE for a horse show the day before as well as on show day. Good preparation will help with preshow nerves and allow you to concentrate on having a good ride.

THE DAY BEFORE THE SHOW

> **Tools you will need:** *horse-washing supplies, hair polish, a light sheet or blanket, braiding supplies, bandage for tail, clippers*

☐ **BATHE THE HORSE** (see pages 46–49) and use hair polish to repel stains. Do not put hair polish on the saddle area as it can be slippery. After the horse is dry, put on a light sheet or blanket to keep him clean.

☐ **BRAID THE MANE & TAIL** if riding huntseat English (see chapter 5).

☐ **WRAP THE TAIL** carefully with a bandage to keep it clean and tidy overnight. Wrap tightly enough to keep it on, but not so tight that it cuts off the circulation to the tail.

☐ **CLIP WHISKERS, BRIDLE PATH, JAW LINE, FELTLOCKS & THE EDGES OF THE EARS** (see pages 44–45). If your discipline requires that all hair be clipped out of the ears, consider asking a professional to help, since many horses object to this. Afterward, carefully clean the inside of the ears with some alcohol on the corner of a rag.

THE DAY OF THE SHOW

Tools you will need: *shampoo, bucket & sponge, corn starch, banding supplies if riding Western, grooming & hot towel supplies, hair polish, baby oil*

☐ **WASH SOCKS** if the horse has them. Apply corn starch to white socks while they are still damp, wait for the corn starch to dry, and brush with a soft brush.

☐ **WESTERN BANDING** (see page 54) should be done in the morning before the class.

☐ **CURRY, THEN HOT-TOWEL** with a small amount of hair polish mixed in the water.

☐ **BRUSH THE HORSE** (see pages 36–39), **PICK OUT HIS FEET** (see pages 42–43), **& COMB OUT HIS TAIL.**

☐ **CLEAN NOSE & MUZZLE** with a sponge or the hot towel.

JUST BEFORE THE CLASS

Tools you will need: *hoof oil, rag or towel, baby oil.*

☐ **PAINT THE FEET** with hoof oil or use a commercial hoof polish (see page 43).

☐ **POLISH THE HORSE'S COAT** using a towel or rag.

☐ **RUB A SMALL AMOUNT OF BABY OIL** on your hands and apply it to the muzzle, the inside of the ears, and lightly around the eyes. Use what is left on your hands to comb through the long hairs of the tail with your fingers.

Basic Horse Anatomy

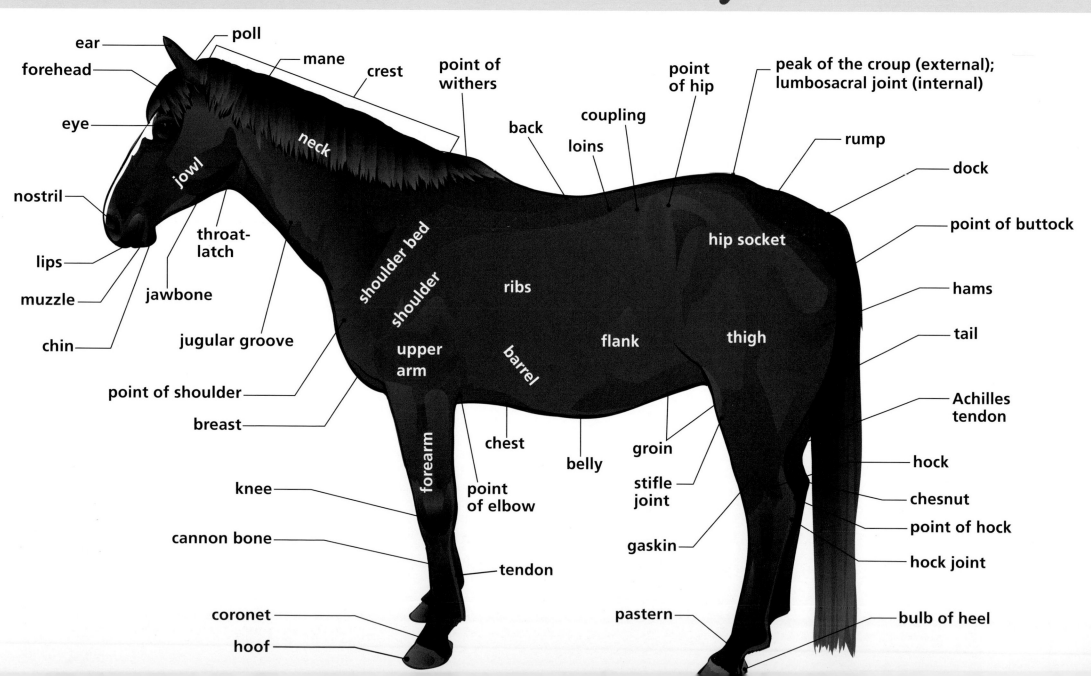

ear

poll

forehead

mane

crest

eye

point of withers

point of hip

peak of the croup (external); lumbosacral joint (internal)

neck

back

coupling

rump

jowl

loins

nostril

dock

throat-latch

shoulder bed

point of buttock

lips

hip socket

muzzle

jawbone

shoulder

ribs

hams

chin

jugular groove

flank

thigh

tail

point of shoulder

upper arm

barrel

Achilles tendon

breast

chest

groin

hock

forearm

belly

stifle joint

chesnut

knee

point of elbow

point of hock

cannon bone

gaskin

hock joint

tendon

coronet

pastern

bulb of heel

hoof

Glossary

BARN SOUR. A bad habit that may result in a horse bolting back to the barn or to his herd-mates.

BILLET. A leather piece that attaches the girth to the saddle.

BRIDLE. Headgear for controlling the horse, consisting of headstall, bit, and reins.

BROW BAND. A leather strap of bridle that goes around the horse's forehead.

CAVESSON. A noseband attached to a headstall.

CHECK. To contact the mouth with the bit; to cue the horse to slow or stop or pay attention.

CINCH RING. A metal ring on the end of a cinch that fastens to the saddle billet or latigo.

CROSS-TIE. A means of tying a horse in which a chain or rope is attached to each side ring of the horse's halter.

CURB. A bit with shanks. It puts pressure on the mouth, chin groove, and top of head when the reins are pulled.

CURB STRAP. Fastened to the top part of a curb bit, this puts pressure on the chin groove.

ERGOT. A small, horny structure located at the back of the fetlock joint.

FROG. The thick, triangular-shaped tissue on the bottom of a horse's hooves.

GIRTH. A strap under the belly to hold an English saddle in place. Also, the body area of the horse where this goes.

HAND. A unit of measurement used to measure from the highest point of the withers to the ground. One hand equals four inches.

HEADSTALL. A headpiece that holds a bit in place.

LATIGO. A leather strap that fastens the cinch to the saddle.

LONGE. To work a horse in a circle usually on a 30-foot line around you at various gaits.

NEAR SIDE. The horse's left side.

OFF SIDE. The horse's right side.

RIGGING. Straps connecting the cinch to the saddle.

SHEATH. The skin folds that encase a horse's penis.

SNAFFLE. A bit with rings attached at the mouthpiece. Pulling on the reins puts pressure on the tongue, bars, and mouth corners.

TAIL RUBBING. A habit that may originate from anal or skin itch or a dirty sheath or udder.

TWITCH. A means of restraint. A nose twitch is often a wooden handle with a loop of chain, applied to the horse's upper lip.

Emergency Contact Numbers

POST COPIES OF THIS INFORMATION in an easily accessible place in your barn, home, and trailer. If you are traveling with your horse, it's a good idea to also tuck a copy into your suitcase or bag.

PRIMARY VETERINARIAN

Name:	Alternate phone:
Office phone:	Cell phone:

SECONDARY OR EMERGENCY VETERINARIAN

Name:	Alternate phone:
Office phone:	Cell phone:

LOCAL BACKUPS (family, friends, neighbors, other local horsepeople)

Name:	Number:
Name:	Number:
Name:	Number:

EMERGENCY NUMBERS (In most areas, dial 911 for emergency assistance.)

Fire department:
Insurance company:

SUPPLIERS

Feed supplier:
Tack and equipment supplier:
Trailering company:

OTHER NUMBERS

Barn:	Truck and Trailer:
Home:	Cell phone (if applicable):